# WORDS YOU SHOULD KNOW: 2013

## The 201 Words from Science, Politics, Technology, and Pop Culture That Will Change Your Life This Year

Nicole Cammorata

Avon, Massachusetts

Published by
Adams Media, a division of F+W Media, Inc.
57 Littlefield Street, Avon, MA 02322. U.S.A.
*www.adamsmedia.com*

ISBN 10: 1-4405-5640-7
ISBN 13: 978-1-4405-5640-1
eISBN 10: 1-4405-5641-5
eISBN 13: 978-1-4405-5641-8

Printed in the United States of America.

10   9   8   7   6   5   4   3   2   1

This publication is designed to provide accurate and authoritative information with regard to the subject matter covered. It is sold with the understanding that the publisher is not engaged in rendering legal, accounting, or other professional advice. If legal advice or other expert assistance is required, the services of a competent professional person should be sought.

—From a *Declaration of Principles* jointly adopted by a Committee of the American Bar Association and a Committee of Publishers and Associations

Many of the designations used by manufacturers and sellers to distinguish their product are claimed as trademarks. Where those designations appear in this book and F+W Media was aware of a trademark claim, the designations have been printed with initial capital letters.

*This book is available at quantity discounts for bulk purchases.*
*For information, please call 1-800-289-0963.*

For my mother and father, who taught me my first words, and for my brother, who listened.

For Karla, who sees the humor in words.

And for John Hopkins, who somehow knew I'd write this book long before I did.

# contents

# Introduction

*Geoengineering. Craftermath. Smart clothes.*

Do you know what these words mean? If not, you'll need to in 2013. While new discoveries and developments happen all the time, experts expect there to be more progress this year than in any previous year due to our advanced technology, social media, and the ever-changing face of pop culture. The fast-moving pace of our lives today makes it all the more important for you to get a grasp of the 201 newsworthy words found throughout this book now so you'll have an idea of the changes that will happen and a better understanding of the headlines that will grace newspapers and impact your life in 2013 and beyond.

From Google's *Project Glass* and *grey divorce* to *Space Adventures*'s proposal for out-of-this-world travel, the terms found throughout these pages are pulled from the minds of well-known experts, innovators, and authorities, and will change the way you see the world in 2013. They are defined in easy-to-read language and are accompanied by a pronunciation guide and a sample sentence that will show you how to correctly use the terms in conversation. For each entry, you'll also find a list of related words, so that you can learn how these concepts will come together by the end of the year.

So what are you waiting for? It's time to learn the words that will dominate discussions—and your life—this year!

# #
# &
# numbers

**#waywire** *(WEY-wahy-ir), noun*

#waywire is an online news site aimed at the Millennial generation, who may not be getting their news from traditional forms of media like newspapers and magazines, or who may feel that these mainstream media sources are neglecting their needs and their voice. Created by a team of Silicon Valley innovators and Newark mayor Cory Booker, the project has received backing from Eric Schmidt of Innovation Endeavors, Josh Kopelman of First Round Capital, and Troy Carter, manager for Lady Gaga. People like Oprah Winfrey and LinkedIn CEO Jeff Weiner are reportedly also investors. A story by E. B. Boyd in Fast Company tells readers to think of #waywire "as a partially user-generated Funny or Die, but for news and social issues, with Pinterest/Twitter-style social dynamics." The site is intended to be both a news stream and a collaborative portal for the Millennial generation, who will be encouraged to react to news by posting their own videos on the site. The name #waywire includes a hashtag, perhaps to encourage sharing of the content on Twitter the way hashtags spark conversations, retweets, and participation on the site. #waywire could very well be the next big social media site that everyone's talking about this year.

**RELATED WORDS:** Facebook, Millennial generation, Twitter

**HOW YOU'LL USE IT:** *"I feel like I only just mastered Twitter and now I hear about this #WAYWIRE thing—I can never keep up!"*

## & other stories *(and UH-ther stohr-eez), noun*

Swedish clothing company H&M is launching a new chain of retail stories this year, called & other stories. While H&M has historically billed itself as selling fashion-forward clothing at affordable prices, its new sister store will up the ante a bit in quality and will be considered a luxury retailer that sells clothes at a higher price point. & other stories will join another upscale offshoot of the H&M brand, the limited-edition retailer COS. COS and H&M have teamed up with designers like Marni, Versace, and Lanvin in the past—it is rumored that & other stories will follow suit and also include limited-edition partnerships with big-name designers. First reported in the fashion news publication *Women's Wear Daily*, a press representative for H&M named Håcan Andersson confirmed the name of the store in the spring of 2012. Though starting the name of a store with an ampersand may be a bit confusing at first, its use in & other stories will help to maintain a brand identity with H&M as it visually links it back to the mother brand. H&M stands for Hennes & Mauritz and has more than 2,300 stores in forty-three countries around the world.

**RELATED WORDS:** COS, H&M, Sweden

**HOW YOU'LL USE IT:** *Is that cute dress you wore to my party from H&M's new store & OTHER STORIES?"*

**2** 

**2012 DA14 asteroid** *(twen-tee TWELVE dee ey fohr-TEEN AS-tuh-roid), noun*

Discovered in 2012 by a team of astronomers at the Observatorio Astronomico de La Sagra in Spain, the 2012 DA14 asteroid is expected to pass extremely close to the Earth this year. Astronomers have predicted that the asteroid will zoom by a mere 17,000 miles from the Earth on February 15, 2013. This may still sound like it's far away to some, but think of it this way: the asteroid will travel between the Earth and the manmade satellites that orbit it. It measures about 150 feet in diameter and has an orbit that resembles that of the Earth. It's definitely on the smaller side, but even an impact with something that small could have big consequences. Although even if it did hit the planet, there's a 70 percent chance it would land in the ocean. In discussing the asteroid, astronomers have been adamant that 2012 DA14 will not collide with the Earth during this visit. However, future visits could pose a great possibility that it would hit. After its trip past Earth in 2013, the next visit from the asteroid is expected in 2020.

**RELATED WORDS:** doomsday preppers, Palermo Technical Impact Hazard Scale, Tunguska event

**HOW YOU'LL USE IT:** *"Even if the 2012 DA14 ASTEROID doesn't hit us, I'm still pretty sure that, someday, an asteroid will be the thing that does us in—just like the dinosaurs."*

## 3-D printing *(three-dee PRIN-ting), noun*

This year is poised to be the year of 3-D printing as a plethora of sectors get in on this emerging technology. 3-D printing creates a three-dimensional object, and the technique can be applied to a host of different fields. The way 3-D printing works is actually pretty simple: input a design and a special printer will create it one layer at a time. The amount of materials available to 3-D printers—things like plastic, ceramics, glass, and steel—is growing. Researchers have already had success incorporating the technology into the biomedical field, creating bioartificial bladders using 3-D printing technology by layering cells on top of each other. 3-D printing is also being used to create clothing—most notably by design duo Mary Huang and Jenna Fizel of Continuum Fashion. Areas like engineering, geography, aerospace, and automotive are also utilizing the technology. 3-D printing has been around for the past decade, but a group of new applications as well as at-home hobbyist kits have thrust it into the limelight for 2013. Among the things being printed in this way are musical instruments, shoes, clothing, jewelry, and even ammunition and brass knuckles. Law enforcement officials have expressed concern that, down the road, as the technology becomes more mainstream, it could create a situation where anyone with a 3-D printer could create things like weapons, or keys, at will.

**RELATED WORDS:** additive manufacturing, bioartificial

**HOW YOU'LL USE IT:** *"I'd love to use 3-D PRINTING at home so that I could just print out a new necklace or a pair of earrings in the morning when I'm getting dressed."*

4

**4-D** *(fohr-dee), noun*

In the past few years, moviegoers have grown accustomed to 3-D viewing experiences where special glasses allow viewers to see images that appear to leap out at them from the screen. The experiences will go a step further in the coming years when 4-D movie theaters become mainstream. A 4-D movie experience is completely immersive, enveloping the viewer in an environment meant to mimic what's happening on screen. This means that when there's an explosion happening in the movie, the viewer's seat will move or vibrate, simulating the feeling of what they're seeing. A 4-D movie may also feature mist, fans, strobe lights, and even various scents pumped through the theater, known as aromascope or smell-o-vision. The concept has been around for some time now, with movies in the 1950s featuring vibrating seats and the 1960 smell-o-vision flick *Scent of Mystery*. In 2011, *Spy Kids 4: All the Time in the World* included scratch-and-sniff cards to accompany various scenes during the movie, and East Coast home retailer Jordan's Furniture's MOM—motion odyssey movie—has been around since 1992. The 4-D movie technology is about to become more mainstream as the trend is growing in places like South Korea, Thailand, and Mexico. At the forefront of this technology is the South Korean company CJ Group, which has plans to open two-hundred 4-D theaters in the United States in the next five years.

**RELATED WORDS:** 3-D, aromascope, smell-o-vision

**HOW YOU'LL USE IT:** *"I hear that new tornado movie is supposed to be awesome in 4-D."*

# A
# B

### Advanced LIGO *(ad–vanst LAHY-goh)*, *noun*

**A**

In much the same way that the Higgs Boson became a defining term for the physics and astrophysics communities in 2012, LIGO could very well help to create a defining moment for these fields in the coming years. LIGO, which stands for Laser Interferometer Gravitational-Wave Observatory, is a multimillion dollar physics project cofounded by teams at MIT, Caltech, and more in 1992. The project, which is funded by the National Science Foundation, has largely been to search for and determine the existence of cosmic gravitational waves, a theoretical occurrence that scientists hope to confirm. The program is getting an upgrade this year. A new, more sensitive detector called Advanced LIGO is expected to be developed and will become fully operational by 2014. According to an MIT website for the project, "the Advanced LIGO project will completely upgrade the three U.S. gravitational wave interferometers, bringing these instruments to sensitivities that should make gravitational wave detections a routine occurrence." It is widely believed that Advanced LIGO may help to advance cosmic gravitational waves from theory to fact, study black holes, and further confirm Albert Einstein's theory of relativity.

**RELATED WORDS:** cosmic gravitational waves, theory of relativity

**HOW YOU'LL USE IT:** *"I heard someone talking on the train the other day about ADVANCED LIGO and how the physics community is probably going to make a big announcement about their findings soon."*

**ADzero** *(ad-ZEER-oh), noun*

Who says the latest technology can't be environmentally friendly too? Designed by Middlesex University student Kieron-Scott Woodhouse, ADzero is an Android smartphone made out of bamboo. The mission of the phone—and the ensuing design company—is "Reinventing Simplicity." The body of the phone is crafted out of a single piece of organically grown bamboo wood, a renewable resource, while the internal mechanism of the device has been made using as many recyclable materials as possible.

"Bamboo may seem like a strange material to use for a phone, but it's actually extremely strong and very durable, perfect qualities for this kind of application," Woodhouse said, as reported in Britain's the *Telegraph*. The phone is said to weigh half of what an iPhone does. The ADzero phone will first launch in the UK and China, then hopefully trickle down to the United States. When the phone does finally hit the market, it will also include new smartphone camera technology called "ring flash." Built directly into the wooden casing, the flash produces more even lighting than traditional smartphone cameras. The launch of an eco-friendly smartphone comes at a time when e-waste—discarded electronics like computers, microwaves, and mobile phones—is on the rise. Many states have banned e-waste from landfills, with Colorado joining the pack this year.

**RELATED WORDS:** e-waste, ring flash

**HOW YOU'LL USE IT:** *"I'm thinking about getting one of those new ADZERO phones. Since each one is made out of a piece of bamboo, no two are exactly alike."*

**A**

**aerostats** *(AIR-oh-stats), noun*

United States involvement in Iraq and Afghanistan has brought about the return of aerostats, blimp-like spy crafts last used in the 1960s. Back in 1998, the U.S. Army worked with American defense contractor Raytheon to develop an airborne missile-detecting sensor that could be kept tethered to the ground. Called JLENS, the Joint Land Attack Cruise Missile Defense Elevated Netted Sensor system allowed the military to better detect low-flying cruise missiles. A few years later, the ships were adapted so that they could be used to survey surrounding areas visually, not just with radar. "Known as Rapid Aerostat Initial Deployment, or RAID, the static blimps soon became a signature of U.S. bases in Iraq and Afghanistan," wrote David Axe in an article about military airships on AOL.com. At their simplest definition, aerostats are spy balloons, and they're a major threat against insurgents in the Middle East. A May 2012 article in *Newsweek* described aerostats thusly: "More than 100 feet long and packed with surveillance cameras and electronic eavesdropping equipment, the airships float tethered above cities, U.S. bases, and strategic highways." Expect to hear mention of aerostats as the U.S. military sorts out the future of these high-tech spy crafts. For their 2013 budget, the Pentagon proposed cost-saving measures that would include grounding twenty-eight aerostats, saving some $6 billion.

**RELATED WORDS:** JLENS

**HOW YOU'LL USE IT:** *"The element of surprise is back now that the U.S. military has grounded all those AEROSTATS."*

**Aha** *(ah-HA), noun*

A

The simple car radio will be getting a serious upgrade this year when the new Suburu BRZ debuts with Aha as an option. To help attract "younger and tech-minded buyers, what was once a humble car radio is being packed with increasingly complex features that connect owners and their gear to their cars and to each other," writes Jim Gorzelany in a *Forbes* article titled "Hottest New-Car Features for 2013." The Aha system will allow drivers to access all of the media and information on their smartphones, something that can be augmented as time goes on. Aha radio assembles favorite radio stations, podcasts, music, even traffic reports, in one central—and mobile—location. Sure drivers can listen to the radio and play music, but they'll also be able to search for a restaurant, get directions, sports scores, you name it. Working in tandem with Pioneer, an app-driven car dashboard, drivers (and passengers) can even have Aha read aloud updates from Facebook and Twitter using text-to-speech technology. Smartphone users who have downloaded the app will be able to stream their data from the phone and into the car using a USB connection or Bluetooth device.

**RELATED WORDS:** infotainment, text-to-speech

**HOW YOU'LL USE IT:** *"So glad my new car has AHA—it makes long car rides way easier."*

**A**

**amazeballs** *(uh–MEYZ-bawlz), adjective*

The word amazeballs has been picking up steam since 2009 when celebrity blogger Perez Hilton successfully made the word a trending topic on the social networking site Twitter. YouTube comedy duo Jessica and Hunter have also been credited with helping to make the term popular. Some have speculated that it will be officially added to the dictionary this year. The word amazeballs is used to connote positive feelings about something and is derived from the work amazing. It is stronger than "that's great" and can be used in the same way you'd use the words awesome, fantastic, and wonderful. The word is frequently coupled with the word totes (short for totally), as in "this sandwich is totes amazeballs." Since the word first began to circulate on social media, its use has become more mainstream and is poised to become one of the hottest slang terms for 2013. Since Perez Hilton got it trending, the word has been used on *Glee*, and various celebrities such as Simon Pegg, Giuliana Rancic, and Kate Walsh have used the word on Twitter. In 2010, Yahoo.com even featured a story called "The Most Amazeballs Women of 2010." Late in 2012, the UK branch of Kellogg's created a box of cereal called Totes Amazeballs at the request of British singer Tim Burgess. There was just one box made, but the company has since fielded numerous requests to produce the super-sweet cereal for the masses.

**RELATED WORDS:** amazing, Perez Hilton, totes, Twitter

**HOW YOU'LL USE IT:** *"How AMAZEBALLS would it be if we ran into some actual celebrities during our trip to Los Angeles?"*

## Anthropocene *(AN-thruh-puh-seen)*, *noun*

Also called the "age of man," the Anthropocene is the epoch you are currently living in and is marked by a rise in $CO_2$ production, the growth of cities, and the rapid consumption of fossil fuels. The things that define this epoch though are the elements that could (and most likely will) destroy it. "Human beings have so altered the planet in just the past century or two that we've ushered in a new epoch: the Anthropocene," writes *Field Notes from a Catastrophe* author Elizabeth Kolbert in an article about the Anthropocene in *National Geographic*. Coined by biologist Eugene F. Stoermer but popularized by Dutch chemist Paul Crutzen, there's still some discussion on whether or not the term can be officially used. Many experts still believe this to be the Holocene period, but as the use of the word Anthropocene increases colloquially, its scientific usage is gaining traction. Determining the true start and finish to an epoch, of course, is somewhat complicated. In the near future—perhaps even this year—we may have the answer as scientists, paleoclimatologists, and experts alike are working on determining the exact start to the Anthropocene. It will be big news once that question is answered, since only then can it be officially included on the geologic timescale. Ultimately, it will be the International Union of Geological Sciences that gets to make the call.

**RELATED WORDS:** industrial revolution, paleoclimatologists, stratigraphers

**HOW YOU'LL USE IT:** *"It could be the end to the ANTHROPOCENE as we know it as soon as the fossil fuels are gone."*

**A**

**anti-austerity movement** *(AN-tahy-aw-STER-i-tee moov-muhnt), noun*

The anti-austerity movement was surging through Europe during the summer of 2012 and is likely to spill over into 2013 as unrest and debt issues in the Eurozone continue. Those who align themselves with the anti-austerity movement are protesting austerity measures by the government meant to cut a country's deficit largely by way of limiting public services. A big portion of anti-austerity protesters includes public-sector workers like firefighters, police officers, education workers, and those related to health care fields, since these are the jobs most affected by government austerity measures and spending cuts. Meanwhile, people like British Parliament member and the Leader of the Labour Party and Leader of the Opposition Ed Miliband and French President François Holland have openly discussed taking an anti-austerity approach to solving the Eurozone crisis. According to a story by Jane Merrick in the *Independent*, Miliband's plans for an anti-austerity approach would "be based on 'measured' deficit reduction and growth policies from those countries enjoying historically low interest rates" rather than targeting those that are already down and out.

**RELATED WORDS:** austerity, Blockupy Movement, Eurozone, Grexit, Occupy Movement

**HOW YOU'LL USE IT:** *"I heard there were more protests in Europe over the weekend related to the ANTI-AUSTERITY MOVEMENT."*

**anticipointment** *(an-tis-uh-POINT-muhnt)*, *noun*

When the promise of a new product, movie, or technological advance doesn't live up to the hype, that's anticipointment. The word is a combination of anticipation and disappointment and its use happens when those two sentiments intersect. Oftentimes it's the media's overwrought hype machine that can almost ensure anticipointed feelings if a product or event has been excessively covered. Though the word has been around for years, this year's new releases could certainly push the word back into the limelight. For one, there's the much awaited followup to Will Ferrell's comedy *Anchorman: The Legend of Ron Burgundy*. The sequel to the 2004 flick, almost ten years in the making, is called *Anchorman 2* and is due to hit the big screen this year. What else could lead to potential anticipointment in 2013? There's the promise of an Apple TV set and the completion of the world's biggest skyscraper, called Sky City, in China. Or maybe it will be with the release of Google's super advanced glasses that can display notifications and content from your phone onto a mini screen just in front of your eyes. What if none of this is as cool as we all hoped? Well, that's just a big anticipointment.

**RELATED WORDS:** Apple iTV, Project Glass, Sky City

**HOW YOU'LL USE IT:** *"I hope the new* Anchorman *sequel doesn't turn out to be a huge ANTICIPOINTMENT—I've been waiting so long for this movie."*

**A**

**Apple iTV** *(ap-uhl AHY-tee-vee)*, **noun**

While Apple TV has allowed for the seamless display of things like movies and TV shows that you have stored in iTunes on your own TV set, the Apple iTV takes things to a whole new level. If/when it launches, this will be the hot technology item for the year. The Apple iTV is rumored to be a television set complete with an HD monitor. In a blog post on the *New York Times* website, Nick Bilton wrote in 2011 that he "learned that executives at Apple knew as far back as 2007 that the company would eventually make a dedicated TV. This realization came shortly after the company released the Apple TV, a box that connects to any manufacturer's television to stream iTunes content. Consumers did not flock to the Apple TV, and rather than abandon the project, Apple began calling it a 'hobby.'" That version of Apple TV is now in its third generation, but will certainly be overshadowed by the more sophisticated Apple iTV. Among the features that the product is rumored to have: a touchscreen monitor or touch screen remote control (like an iPod touch you use to control the TV), motion-sensor technology, and a large online library of content from which to choose.

**RELATED WORDS:** Apple, fanboy/fangirl, iPad, iPhone, iPod, Smart TV

**HOW YOU'LL USE IT:** *"I don't care how much it costs—I just think the APPLE ITV is so cool and I can't wait to buy it."*

**Aquarius Reef Base** *(uh-KWAIR-ee-uhs REEF beys), noun*

An undersea lab owned by the National Oceanic and Atmospheric
Association that's been in use for two decades is at risk of being
shuttered after its funding was cut in 2012. Called the Aquarius
Reef Base—or just Aquarius for short—the lab is located in Key
Largo, Florida, 60 feet below the water. Up to six divers—or
aquanauts, as they are sometimes called—live in Aquarius in a
space that measures just 400 square feet. The underwater lab has
bunk beds, a kitchen, running water, and computers. When these
aquanauts travel to Aquarius, they usually stay for about ten days
at a time while they conduct their research, work on film projects,
and so on. The base allows them to stay at a low depth and conduct
their work without constantly having to return to the surface and
thus risk the bends. The bends is a dangerous condition that happens when divers surface too quickly, thus developing nitrogen
bubbles in their muscles. Expect to hear more about the Aquarius
Reef Base as the newly developed Aquarius Foundation tries to
raise the funds necessary to keep it in operation.

**RELATED WORDS:** aquanauts, the bends, NOAA

**HOW YOU'LL USE IT:** *"I've always wanted to be able to do my
research from the AQUARIUS REEF BASE—I hope it gets
the funding it needs so that I can make that dream a reality
someday."*

**A**

**Areva** *(ah-REEV-ah), noun*

French-based company Areva has energy sites worldwide, covering multiple facets related to the energy industry. They are involved with things like mining, chemistry, reactors, nuclear propulsion, fuel recycling, clean energy, and more. This energy conglomerate—one of the biggest nuclear-reactor builders in the world—is sure to make headlines this year if efforts to start construction on a uranium enrichment plant in Idaho begin as planned. Originally, Areva was supposed to start the work in 2012, but held off so that the company could have time to bounce back from low profitability the previous year. Idaho wasn't the only project delayed because of this—Areva's construction efforts in places like northern France, Finland, and Africa were also put on hold. In fact, many nuclear projects by companies across the world hit roadblocks in recent years because of the partial meltdown in Japan at the Fukushima nuclear reactor and the falling price of uranium that ensued. "Anti-nuclear campaigners such as Greenpeace are quick to argue that Areva's woes demonstrate the folly of new investment in nuclear power," writes Rav Casley Gera on clean-energy news site CleanTecnhnica. The $3 billion uranium enrichment plant in Idaho—dubbed Eagle Rock—a project that was granted a loan of $2 billion from the U.S. Department of Energy, is expected to finally commence this year.

**RELATED WORDS:** clean energy, Eagle Rock, nuclear power

**HOW YOU'LL USE IT:** *"I wonder if there will be any Greenpeace protesters when that AREVA project in Idaho finally starts."*

**Army Retention Initiative** *(ahr-mee ri-TEN-shuhn ih-NISH-ee-uh-tiv)*, **noun**

Quite literally, the Army Retention Initiative is an Army directive aimed at figuring out how to retain military personnel after their service contracts end later this year. It is also referred to as the Army Retention Program. With pressure to downsize the military, the initiative's main goal is aimed at retaining quality soldiers—keeping only the best moving forward. In fact, under the Army Retention Initiative, re-enlistment may be denied to soldiers whom commanders consider to be less than stellar. The plan explicitly states that prior to the expiration of a soldier's contract, a commander should communicate where he or she is lacking and what he or she would need to do in order to be able to re-enlist. According to the U.S. Army's website, "Managing the Army's retention program and its assigned objectives is essential to retaining a quality force with joint and expeditionary capabilities. Today's Soldiers possess a wealth of skills and combat experience, and retaining these Soldiers is essential to the quality of the force." Expect to hear more about the Army Retention Initiatives as American troops are withdrawn from Afghanistan and as the military begins to assemble its scaled-back force.

**RELATED WORDS:** Army Retention Program

**HOW YOU'LL USE IT:** *"Everyone in my troop who's planning to re-enlist is really trying to stay on their toes with this ARMY RETENTION INITIATIVE in effect."*

**A**

### -athon *(uh-thon), suffix*

Derived from marathon, the word for a 26.2 mile road race, terms ending in -athon refer to any prolonged activity where endurance—either of mind or body—is necessary. Words like bikeathon, skiathon, jogathon, and danceathon are some good examples, as is the well-known telethon. Most recently is the development of hackathons. A hackathon is a day-long event that gathers computer programmers, software engineers, web designers, and the like as a way to work with programming systems, develop new tools or applications, or to solve a particular problem. While things like bikathons and telethons are meant to serve as a fundraising event, hackathons are meant to encourage innovation and problem solving among a group with a specific skill set. There's really no limit to the creation of more "-athon"-ending words. Need friends to help you decorate your new apartment? Host a paintathon. Catching up with an old friend after years apart? Must be a talkathon. 2013 could bring a host of these new words to the American lexicon as news develops. Could we see the use of the word buildathon during the expedited construction of Sky City in China? We'll just have to wait and see.

**RELATED WORDS:** bikeathon, danceathon, hackathon, jogathon, skiathon, telethon

**HOW YOU'LL USE IT:** *"I'm participating in a bikeATHON this summer to raise money for charity."*

**attachment parenting** *(uh-TACH-muhnt PAIR-uhn-ting), noun*

A May 2012 story by Kate Pickert in *Time* magazine, and its accompanying cover photo featuring a mother breastfeeding her three-year-old son, helped introduce the term attachment parenting and to spark a discussion about just what it is exactly. Attachment parenting is a child-rearing methodology that stresses the importance of things like breastfeeding, co-sleeping, and joint baths, and advocates that the parent and child spend as much time together as possible. It also suggests the use of baby slings and other child-carrying apparati that keep the baby, quite literally, attached to the parent rather than being pushed in a stroller. The theory behind attachment parenting and the message of Attachment Parenting International (API) is that this parenting style will help raise children who are more secure, independent, and empathetic. Critics of attachment parenting maintain that there hasn't been enough conclusive research carried out to support the claims that this style would be any better than the mainstream. William Sears and his 1993 guide *The Baby Book* are credited with first publicizing the parenting philosophy but it was the *Time* magazine cover story that has turned it into a cultural talking point.

**RELATED WORDS:** breastfeeding, co-sleeping, helicopter parent

**HOW YOU'LL USE IT:** *"My husband and I are really trying to follow the ATTACHMENT PARENTING model so we're not even going to buy a stroller."*

**B** 

### B612 Sentinel *(bee-siks-twelv SEN-tin-uhl),* **noun**

The B612 Foundation has announced plans to send a solar-orbiting telescope, called the B612 Sentinel, into deep space. The purpose of the Sentinel Mission, as it's being called, is to spend the next hundred years charting the orbits of asteroids in our solar system. The name B612 is lifted from Antoine de Saint-Exupéry's French children's book, *Le Petit Prince.* In the book, a little boy lives on a fictional asteroid named B612. Though the project is still a few years off—anticipated launch dates are in 2017 or 2018—we'll be hearing about the B612 Sentinel as the privately operated project gains funding. "B612 marks something totally novel for the private space industry," wrote Clay Dillow in *Popular Science.* "It isn't a commercial venture. It's something more like space philanthropy." The end goal of the Sentinel Mission and the B612 foundation is to be able to better predict and identify which asteroids are on a course for Earth, or at least could be. Scientists have worked for years on how to combat an asteroid that's on a collision course with Earth—this will better prepare us for those moments. The B612 Sentinel will carry an infrared telescope, scanning the sky for moving objects with the hope that within five-and-a-half years it will have mapped about 98 percent of near-Earth asteroids.

**RELATED WORDS:** 2012 DA14 asteroid, GAIA mission, gravity tractor, Sentinel Mission, space philanthropy, Space X

**HOW YOU'LL USE IT:** *"We can all sleep a little better knowing that the B612 SENTINEL is going to help us determine which asteroids are a threat before they're actually a threat."*

**babesiosis** *(buh-beez-e-OH-sis), noun*

Health officials are warning people of a tick-borne disease called babesiosis that we're only just starting to study more here in the United States. Though it's been around for years, many people don't know about it because it doesn't receive the same kind of coverage and media attention that Lyme disease does. Named for Dr. Victor Babes, who first identified the disease in cattle back in 1988, babesiosis was not discovered in humans until 1957. A second case, documented on Nantucket island in 1969, helped the disease to earn the nickname Nantucket fever. A recent study of the disease in Rhode Island is helping to bring discussions of babesiosis to the forefront of conversations, with some predicting that it may soon rival Lyme disease in the United States. You get babesiosis the same way you would Lyme—from a tick bite—and the symptoms, things like fatigue, fever, chills, muscle and joint pain, and nausea, resemble malaria. It is most common in the United States, specifically in the Northeast and Upper Midwest, and is spread via mice and deer. In some people, babesiosis may simply clear up on its own, while others require antibiotics.

**RELATED WORDS:** Lyme disease, malaria, Nantucket fever, ticks

**HOW YOU'LL USE IT:** *"I have to remember to pack the bug spray with DEET in it when we go camping so that we're protected from BABESIOSIS and Lyme disease."*

**babymoon** *(BEY-bee-moon), noun*

Before the birth of their first child, many couples are engaging in a short trip, often to somewhere tropical, so they can savor the quiet simplicity of life before kids. It's a time to reconnect with each other before caring for the baby becomes everyone's top priority. A derivation of the post-wedding honeymoon, a babymoon may also feature indulgent spa treatments and lots of relaxation. The original use of the term meant the time period directly after the baby's birth, in which the new parents bond with the baby and remain in a happy, trance-like state, "mooning" over their little one. Author Sheila Kitzinger is credited with coining the term babymoon in her book *The Year after Childbirth*. Thank celebrities and magazines for reinvigorating the term and reclaiming it to mean the prebirth vacation. "Just wrapped up an incredible babymoon at Jade Mountain," tweeted Nick Lachey after a trip to St. Lucia with his pregnant wife, Vanessa Minnillo. Resorts and hotels are now offering special babymoon packages for 2013 travel as a way to lure in customers, and sites like babymoonfinder .com and babymoonguide.com have sprung up as the word—and concept—continues to trickle down through the masses.

**RELATED WORDS:** babymooning, babymoons

**HOW YOU'LL USE IT:** *"Amy and Scott seem so relaxed since they got back from Hawaii—they really have that BABYMOON glow."*

## balkanize *(BAWL-kuh-nahyz), verb*

**B**

The word balkanize (which can also be spelled balkanise) has long been used as a geopolitical term, meaning to break a region apart into smaller states. It comes from the late-nineteenth-century division of the Balkan Peninsula. These days, the term is gaining traction as people use it to summarize their fears about the future of the Internet. "The Internet has been a great unifier of people, companies, and online networks," read the subhead for a story in the *Economist* in 2010. "Powerful forces are threatening to balkanise it." The article describes the Internet's inception, saying that at its outset, it was like the Wild West. There were no rules, no restrictions. For years, anyone with a signal could access whatever they wanted, could make it do whatever they wanted. Now, though, that environment has changed, as governments seek to regulate content on the web and as companies restrict their products to their devices. (Apple's app store is a good example of this.) There are browsing restrictions that occur when traveling overseas, as any American who was tried (and failed) to watch Hulu or access Spotify in Europe can attest. Most notably, of course, is the widespread censorship of content in China. Widespread balkanization of the Internet means more restrictions like this, more closed systems that limit what you can do based on where you are, what device or service you're using, and what you're trying to find.

**RELATED WORDS:** Hulu, iTunes

**HOW YOU'LL USE IT:** *"I'm so sick of companies trying to BALKANIZE the Internet and set limits and restrictions on everything."*

**B**

**beervana** *(beer-VAH-nuh), noun*

America's love for craft beer is on the rise, a trend that has been on a steady increase during the course of the last few years. This brew culture has created the perfect environment for the development of beervanas—areas where breweries have become concentrated and superfluous. The word itself is a combination of the terms "beer," that alcoholic beverage made from grains and hops, and "nirvana," a place of enlightenment and a state of bliss. The word implies that where there is beer, there is happiness. A clever infographic created by designers at Pop Chart Lab features a map detailing all of the breweries in the United States. In areas where the concentration is most dense, there are zoomed-in bubbles allowing for a closer look. "Because breweries cluster in certain 'beervanas' like Portland, we decided to use insets to show detail," the map's creators told Mark Wilson for a story that ran in Co.Design, an online publication of Fast Company. The term beervana originated in Portland, where it also doubles as a nickname for the Oregon city due in part to its ideal growing conditions as well as its large number of breweries—more than anywhere else in the world. Oregon's public broadcasting network detailed this brew culture in their TV program titled *Oregon Experience: Beervana,* which traced the history of beer making in the city from early settlers to present day.

**RELATED WORDS:** hopheads

**HOW YOU'LL USE IT:** *"Anyone want to take a BEERVANA road trip across the country from Boston to Portland?"*

**Belviq** *(bel-VEEK)*, **noun**

When the diet pill Belviq was approved by the Food and Drug Administration on June 27, 2012, it became the first pill of its kind to win such backing in thirteen years. (The last long-term diet pill to be approved by the FDA was Xenical in 1999.) The drug approval comes at a key time, when obesity rates and the number of people suffering from type-2 diabetes are quite high. While diet and exercise has been the long-prescribed method of weight loss, a drug such as Belviq is said to bridge the "treatment gap" that exists between traditional weight-loss measures and something more extreme, like surgery. Belviq, which has also been known by the name lorcaserin, was developed by the San Diego–based Arena Pharmaceuticals and will be sold by Eisai Inc. It is meant specifically for those who are considered to be obese (that means a BMI of 30 or more) or those who are overweight and also suffer from things like high blood pressure, high cholesterol, or type-2 diabetes. According to Rob Stein's post on NPR's health blog, Belviq is "a twice-a-day pill that suppresses appetite and appears to affect metabolism by influencing levels of the brain chemical serotonin." The drug was actually first rejected by the FDA back in 2012 but after submitting new data to alleviate concerns about potential side effects, Belviq received approval.

**RELATED WORDS:** Merida, Orlistat, Qnexa, Xenical

**HOW YOU'LL USE IT:** *"Diet and exercise just isn't working for me—I'm thinking I might ask my doctor about that new BELVIQ that's coming out."*

**B**

## Big Data *(big DA-tuh), noun*

The term Big Data describes the massive amount of information that is created, collected, and catalogued every day; something that is becoming exponentially larger thanks to the Internet and other sophisticated technologies. The archiving of information is a good example of Big Data, from photos and videos to medical records and information associated with social media networks, Internet search histories, and other social data. The professionals that are dealing with Big Data are many, including members of the military, biologists, meteorologists, and astronomers. Technological advances have opened us up to a wealth of data and information, which in turn has allowed us to analyze trends, make predictions, and tell stories like never before. Determining how to manage Big Data will be a hot topic for 2013 and could even lead to job creation as experts in the field are beginning to develop. A 2012 story in the *New York Times* by Steve Lohr called "The Age of Big Data" claimed that the industry of technology analysts is growing as we try to keep up with the creation of Big Data. As it grows, the need for people with special skills suited to how best to manage and analyze all this information is growing, too. It is estimated that the data we're collecting doubles every two years.

**RELATED WORDS:** Industrial Internet, Internet of Things, petabytes, terrabytes

**HOW YOU'LL USE IT:** *"For numbers-loving people who like charts and graphs and making sense out of large amounts of information, BIG DATA is going to be the field to get into in the coming years."*

**bioartificial** *(BAHY-oh-ahr-tuh-FISH-uhl), adjective*

B

The word bioartificial can be a bit misleading at first—or at the very least confusing—since it seems to combine two opposing ideas to produce something that at first glance seems oxymoronic. The prefix bio, of course, refers to something alive—a living thing—while artificial means fake, or manmade. The joining of the two words here works to mean a living thing that has been created in an artificial, and thus unnatural, environment. With the number of patients in need of organ transplants far surpassing the quantity of organs available, there's a new option in development that could possibly close the gap in coming years. Bioartificial organs are those that have been custom-grown using the patients' own cells. Researchers have had some success already in growing bioartificial bladders, but denser, more complex organs like kidneys or livers are proving a bit more challenging. Another method used in bioartificial organ development is to take a donated organ that has been stripped of its original tissue and then apply the patient's own stem cells to the basic structure—called a "scaffold"—allowing it to develop into a working organ. Success of this procedure means that the patient is far less likely to reject the organ, plus it significantly cuts down on the waiting time usually required during organ donation.

**RELATED WORDS:** tissue engineering

**HOW YOU'LL USE IT:** *"I hope that by the time my organs start to give out that scientists will have this whole BIOARTIFICIAL practice perfected."*

**B**

**biometric ATMs** *(bahy-oh-MEH-trik ey-tee-ems)*, *noun*

Biometrics, or biometric authentication, uses sophisticated scanning technology to identify a person as him- or herself and thereby grant control to personal data. Forgoing the traditional ATM card, a biometric ATM scans a piece of the user's identity, most commonly a palm, finger, or retina—something that is both unique to that person and universal to all people. The prevailing thought behind biometrics is that it makes it incredibly difficult to access another person's information or steal their identity. Trouble with debit card theft or loss—especially during a natural disaster—becomes obsolete with the use of these ATMs. Biometric technologies, which may seem like some futuristic idea dreamed up in Hollywood, are becoming more widespread. In Japan, Ogaki Kyoritsu Bank is introducing biometric-only machines that operate by scanning a user's palm, nary an ATM card in sight. By 2013, the India-based Punjab National Bank is planning to install 100,000 biometric ATMs around the country. In Russia, there are even biometric credit-card application machines that assess a user's eligibility using data from the sound of their voice. (The machine can detect changes in the person's speech that indicate whether he or she is lying.)

**RELATED WORDS:** iris scans, next-generation identification

**HOW YOU'LL USE IT:** *"I wish they'd bring BIOMETRIC ATMS to the United States—they just seem way safer than carrying around an ATM card all the time. Plus it's really cool."*

**B**

**BMW i** *(bee-em-DUHB-uhl-yoo AHY), noun*

This year production will begin on vehicles under the newly launched BMW i, a sub-brand of the luxury automaker BMW. The main focus of BMW i will be on eco-friendly and low-emission vehicles. The BMW i3, which is expected to be available this year, is a small, electric car meant to be driven by city dwellers whose space is limited. It will be BMW's first mass-produced vehicle to have zero emissions. The car's internal structure will be made mostly from carbon-fiber reinforced plastic and will seat four. The interior is also expected to be eco-friendly, featuring a dashboard made from eucalyptus wood and seat covers made from wool and leather that has been tanned using olive leaves. Some have speculated that the BMW i3 will even feature a single pedal, which is to act as both the accelerator and the brake. When there isn't any pressure being applied to the pedal, the energy produced by the moving car will recharge the battery. The BMW i3 was previously referred to as the Megacity car or the Mega City vehicle. Following the i3 into production will be a hybrid model named the i8, with three additional models coming down the line from BMW i in the following years.

**RELATED WORDS:** BMW, eco-friendly, GreenGT H2, zero emissions

**HOW YOU'LL USE IT:** *"So who wants to head down to the dealership with me and test out that new car from BMW i?"*

**B**

## brandalism *(BRAN-duhl-iz-uhm)*, *noun*

Brandalism is a new movement among socially conscious street artists that first started to gain traction after a calculated series of street art was successfully completed in the UK in 2012. The word brandalism—a combination of brand and vandalism—seems to have roots in Anne Elizabeth Moore's 2007 book *Unmarketable: Brandalism, Copyfighting, Mocketing, and the Erosion of Integrity.* A widespread brandalism campaign across the UK in 2012 that was eight years in the making, and the media's coverage of the so-called "anti-ads," has helped catapult the concept and the term into conversations. According to a story by Joe Berkowitz on the website for *Fast Company*, "artists were invited to contribute pieces on different themes including Art of Propaganda, Body Image and Well Being, Creative Resistance, Cultural Values, Debt and Environment, Advertising and Visual Pollution." The timing of the 2012 brandalism campaign in the UK was to coincide with the weeks leading up to the summer Olympics in London. (Brandalism is also the name for the community of artists who banded together to carry out this project.) Brandalism involves super-stylized, subversive graffiti, street art, and posters created on top of or in conjunction with preexisting billboards or advertising. The movement has also been referred to as subvertising, due to the kind of anti-consumerism messages it creates.

**RELATED WORDS:** Banksy, gladvertising, graffiti, subvertising

**HOW YOU'LL USE IT:** *"I heard some guys talking on the train the other day about how they're trying to organize some BRANDALISM efforts in the city."*

**breakover** *(BREYK-oh-ver), noun*

The breakover—a post-breakup makeover—occurs when the end of a relationship triggers a new look. It could be a new haircut, a new wardrobe, or anything along those lines that jumpstarts a person out of the depressing slump of being newly single. The term is often used in women's lifestyle magazines or in celebrity news coverage, particularly when a high-profile star changes her look after a breakup. The breakover is now evolving to refer to the reinvention of oneself following any big life change, from bankruptcy to getting fired. A breakover is about seizing the situation and using the opportunity to fuel a change in one's life. The late Nora Ephron hosted a video series on the *Huffington Post* that directly covered this trend. A post on the website explaining the series called people to action thusly: "We're asking you, our readers, to tell us about the hard times, specifically a moment in your life when you were forced to reconsider everything. We call that moment 'The Breakover.'" Not to be confused with the newspaper term, in which a breakover refers to a story that is continued on a separate page.

**RELATED WORDS:** makeover

**HOW YOU'LL USE IT:** *"Reese Witherspoon really looked fantastic after her BREAKOVER following her divorce from Ryan Phillippe."*

**B**

**bridesman/groomsmaid** *(BRAHYDZ-muhn/GROOMZ-meyd),* *noun*

The legalization of gay marriage is gaining traction across the world, with France announcing its plans to make same-sex marriage legal this year. As such, the rise in same-sex weddings means that the traditional roles (and words) for the bridesmaid and groomsman just aren't cutting it anymore. Though the concept of a groomsmaid and bridesman have been bubbling up for years, they're about to hit the mainstream. By definition, a groomsmaid is a female attendant to the groom, while a bridesman is a male attendant to the bride. This trend isn't exclusive to gay weddings, of course, and has been on the rise in the last few years as more couples ditch the traditional "boys on one side, girls on the other" mode of thinking and opt for a more modern interpretation of the bridal party. The challenge for many a bride-to-be, as documented on wedding message boards, is how to dress the groomsmaid or bridesman, and whether or not the groomsmaid should match the boys or the girls. Could this conundrum be the latest fashion emergency to be tackled by bridal magazines in 2013? Only time will tell.

**RELATED WORDS:** groomslady, groomswoman, man of honor

**HOW YOU'LL USE IT:** *"My brother and I are really close so he's going to stand next to me instead of my fiancé and be my BRIDESMAN."*

# C
# D
# E
# F

## C

**Candy Bar Calorie Max** *(KAN-dee bahr KAL-uh-ree maks)*, *noun*

Super-sized candy bars will likely be a thing of the past as of this year. In an effort to combat the obesity epidemic in the United States, popular candy company Mars is halting production on their king-sized candy bars and reducing the calorie count of others. A statement on the company's website reads, "We have committed not to ship any chocolate products that exceed 250 calories per portion by the end of 2013." (Thus the "calorie max.") This means that the regular-size Snickers bar, which is 280 calories, will be getting a makeover. In place of the king-size, which Mars has already begun to phase out, the candy maker will create an option that contains multiple servings of smaller portions to promote sharing. Though Mars plans to cut calories simply by reducing the size of their products, there's been no talk of changing the ingredients. Mars is just one of many companies launching plans to reduce the calorie counts of their products. The initiative is in conjunction with the Partnership for a Healthier America program, a group which has found support in first lady Michelle Obama and whose main goal is to stop childhood obesity.

**RELATED WORDS:** Belviq, obesity

**HOW YOU'LL USE IT:** *"No more king-sized Snickers bars at the Hawkins house this year—Halloween just isn't the same after the CANDY BAR CALORIE MAX."*

**carbon capture** *(kahr-buhn kap-cher), noun*

Carbon capture is a technology used to help thwart the release of the greenhouse gas CO2. The process most often starts at power plants, where fossil fuels are being burned and large quantities of CO2 are being released into the atmosphere. Carbon capture traps these emissions before they can even hit the air. They are then transported to another location and stored, most often deep underground. There are some big concerns about storing CO2 underground, though, with opponents saying that it is not a safe practice since it could lead to leaks and even cause earthquakes. For proponents of this technology, the practice of carbon capture is attractive because it would allow us to continue using fossil fuels while also reducing the bad side effects of doing so. In Connecticut, the Alstom Power plant has been given a $1 million grant by the U.S. Department of Energy to study and develop coal-capture technologies. The EU announced that a large portion of the funding supplied to green research would go to carbon capture. Abu Dhabi also announced plans for a carbon-capture project this year. Though the technology was first developed in the 1930s, it's a hot-button topic among environmentalists and policy makers as we grapple with how to mitigate the effects of fossil fuel use. In the coming years we will certainly hear about new carbon-capture technologies as the process is further studied and put into practice.

**RELATED WORDS:** fracking, geoengineering, ocean acidifcation

**HOW YOU'LL USE IT:** *"If scientists could figure out a way to use CARBON CAPTURE but store it better, it could very well be the thing we've been looking for to help reduce the release of greenhouse gases."*

**C**

**cash mob** *(KASH mahb), noun*

A spin-off of a flash mob, which is when a group of people show up in a public place to perform a choreographed routine or dance, a cash mob is all about spending money and stimulating the local economy. Cash-mob organizers will determine a location—usually based on the votes of group members—and then on a particular day the shoppers will descend upon an establishment with the idea that they will all spend, at minimum, a collectively agreed-upon sum. The amount could be as little as $10 or $20 per person. So as to alleviate stress on shop owners, cash-mob organizers will give them a heads-up so they can prepare for a crowd. The point of a cash mob is to support local businesses during a depressed economy. The name given to a person participating in a cash mob: a mobster. Blogger and engineer Chris Smith is reported to be the originator of the cash-mob phenomenon, which started in Buffalo, New York, in 2011. Since then, the trend has continued to spread and is gaining momentum as local chapters are springing up worldwide and organizing themselves on Facebook and Twitter. Another variation on the theme, called a meal mob, involves dining at a specific restaurant.

**RELATED WORDS:** flash mob, meal mob, mobster, recession

**HOW YOU'LL USE IT:** *"Hey everyone, CASH MOB at Sunshine Lucy's this weekend—who's with me?"*

## cashtags *(CASH-tagz), noun*

Cashtags are an extension of hashtag capabilities on the popular micro-blogging social-media website Twitter. A cashtag makes it possible to hyperlink a stock quote, pointing a user to other Twitter coverage and references to that same cashtag the way a hashtag does. While hashtagging requires you to precede a word with the # symbol, a cashtag precedes a ticker symbol with a dollar sign. Thus, $AAPL would point you to all of the other stock news about Apple, while $GOOG does the same for Google. When the capability for cashtagging was officially launched at the end of July in 2012, there was a slight hiccup. A four-year-old company called StockTwits already had the functionally, and had been curating these cashtags for a few years. StockTwits cofounder and CEO Howard Lindzon likened the cashtag debut to being "hijacked." He wrote in a blog post at the time, "You can hijack a plane but it does not mean you know how to fly it." Expect to hear more about cashtags this year as their use catches on and as the discussion continues about Twitter vs. StockTwits.

**RELATED WORDS:** hashtags, NASDAQ, social media, stock symbol, Twitter

**HOW YOU'LL USE IT:** *"The day my company went public, our CASHTAG almost immediately started trending on Twitter."*

**C**

**Chagas disease** *(SHAH-guhs dih-zeez), noun*

Though the chronic infection Chagas disease has been around for some 9,000 years, researchers have expressed concern that it could turn into a global pandemic. The disease, which was discovered by a Brazilian doctor named Carlos Ribeiro Justiniano Chagas, is also often referred to as the "AIDS of America" since the drugs to treat it are in short supply. (The comparison to AIDS can be a bit misleading, though, since Chagas disease isn't spread through sex.) Though most prevalent in Central and South American countries, there have also been cases reported in southern Texas. According to The Chagas Foundation, "The infection affects up to 20 million people in Mexico, Central America, and South America, making Chagas disease the highest-impact infectious disease in Latin America." In the United States, some 30,000 people are affected. The disease is transmitted by a parasite called Trypanosoma cruzi, which is spread by blood-sucking insects known commonly as "kissing bugs." They are called this because they are known for biting a person's lips in their sleep. Symptoms of the disease at its most severe include the swelling of organs like the heart and liver. A series of studies testing the best course of treatment for Chagas disease are scheduled to conclude in July 2013.

**RELATED WORDS:** kissing bug disease

**HOW YOU'LL USE IT:** *"I'm thinking about canceling my trip to Mexico because I'm sort of freaked out about getting CHAGAS DISEASE."*

## Chang'e-3 mission *(CHUHN-gaw three)*, **noun**

China's space exploration efforts will continue this year with their third lunar mission, dubbed Chang'e-3. Part of the Chinese Lunar Exploration Program, Chang'e-3 will include a six-wheeled, 100-kilogram lunar rover that will explore the surface of the moon for three months and take measurements on the depth of lunar soil and what comprises it. The name of the mission is taken from the Chinese word given to the goddess of the moon, Chang'e. According to Chinese legend, Chang'e and her husband had plans to drink an elixir that would keep them immortal, and thus together forever. Their plans went awry when a man killed Chang'e's husband and threatened her life. Chang'e drank the elixir to escape the man and thus became immortal herself. She decided to live on the moon since it was the closest thing to Earth in the heavens. Thus the Chinese moon festival, held every autumn, celebrates familial bonds. The Chang'e-3 mission will be the first time a craft has explored the moon since 1976, when the Soviet Union's Luna 24 last stopped there. The Chang'e-3 mission comes in the wake of China's efforts to ramp up their space exploration, having become the third country to put someone in space in 2003. China put their first woman into orbit just last year. Looking further ahead, the country also has aspirations to explore Mars and Venus, and to create a permanent space station.

**RELATED WORDS:** Chinese Lunar Exploration Program, Google Lunar X-Prize

**HOW YOU'LL USE IT:** *"I wonder what new things China will find on the moon during the CHANG'E-3 MISSION."*

**C**

## cisgender *(sis-JEN-der), adjective*

Cisgender is essentially the opposite of transgender. If someone who is transgendered identifies with the opposite sex from which they were born, then someone who is cisgendered identifies with the sex written on their birth certificate. "Cis" in Latin means "on the same side as" while gender refers to a person's given sex. Though the word itself is a few years old, use of this word has increased among those who wish to avoid the assertion that a nontransgendered person is "real" or "natural," as is the typical language used when referring to a cisgendered person. "When you have a term for one group of people, you need a word for the rest, and the one they've come up with is 'cisgender.' So the huge majority of us are now, 'members of the cisgender community,'" writes Simon Hoggart in the *Guardian*. Be on the lookout for the use of this word this year, when the Miss Universe pageant will officially expand its contestant pool to include transgendered women in addition to cisgendered women. The pageant, which has been previously limited to cisgendered women, will change the wording in its rules to officially open its doors to transgendered women. The move comes about following the participation of transgendered contestant Jenna Talackova in last year's Miss Universe contest.

**RELATED WORDS:** cis female, cisgendered, cis male, cissexism, transgender

**HOW YOU'LL USE IT:** *"As a CISGENDER myself, I've never doubted that I was supposed to be the sex I was born as."*

## CIVETS *(SIV-its), noun*

CIVETS is the term coined by global forecasting director for the Economist Intelligence Unit Robert Ward to refer to an emerging market comprised of Colombia, Indonesia, Vietnam, Egypt, Turkey, and South Africa. The term CIVETS is an acronym for these countries, which all have populations where the average age is twenty-seven. Skeptics of the term are hesitant to group these countries together in such a way—even in theory—since they see their youthful populations as the only thing uniting these distant countries. Supporters of the term, however, maintain that they are all fast growing with diverse economies and are therefore worth investing in. Supporters also assess that the CIVETS are on track to replace the BRIC, a group of countries comprising Brazil, Russia, India, and China that were deemed worthy of investment nearly ten years ago. Despite the homonym, CIVETS are not, in fact, civets. Civets are nocturnal, cat-like animals native to Southeast Asia known for passing coffee beans through their digestive systems. The coveted beans are said to produce a smoother coffee without an aftertaste.

**RELATED WORDS:** BRICs, civets

**HOW YOU'LL USE IT:** *"I'm going to do my part to invest in the CIVETS this year by traveling to Egypt and Vietnam to support their local economies."*

**C**

**cloud computing** *(KLOUD kuhm-PYOO-ting)*, **noun**

Cloud computing refers to a web-based system of software and data storage that can be accessed by multiple people and alleviates the need of having to reinstall the software on each individual computer. From a user standpoint, all that's required is an Internet connection and a web browser. A good example of this is web-based email like Google's Gmail or Yahoo! Mail. You can access your e-mail from anywhere because it's web-based, allowing you to tap into a central server. Think of Google's suite of web-based products like the calendar or Google Docs. When you interact with your information, it's not being stored locally on your computer, but rather to the cloud. Cloud computing has been on the rise for a couple of years now, slowly picking up steam. Many are predicting a sea change for cloud computing this year, saying that it will take off—especially within corporations, because it's cost-effective and faster than database storage. As David Linthicum predicted on the website for InfoWorld, after the initial ramp-up efforts, there will be a "big wave of implementations in 2013."

**RELATED WORDS:** the cloud, hybrid IT, server

**HOW YOU'LL USE IT:** *"I've become so reliant on CLOUD COMPUTING that I'm not even saving anything to my computer anymore—it's all in Google Docs."*

**connected viewer** *(kuh-NEK-tid VYOO-er)*, **noun**

A connected viewer, which is also sometimes called a connected TV viewer, is a person who is simultaneously watching TV while perusing the Internet on their smartphone, laptop, or tablet device. The reasoning for this multitasking is diverse. A connected viewer might be checking in on social media during commercials, or texting with friends and posting updates and commentary about what they're watching as a way to have a social viewing experience online. The connected viewer may also be searching for information about what they're watching, such as why they recognize a particular actor, the meaning of an obscure reference they just heard, or if something they're watching play out on TV could also happen in real life. This last example is also referred to as the "CSI effect." A 2012 study released by the Pew Research Center found that more than half of all cell-phone users are considered to be connected viewers. The trend is growing and will be something you're talking about—and doing—well into the year.

**RELATED WORDS:** CSI effect, Facebook depression, Facebook phone, fifth screen, FOMO, multitasking

**HOW YOU'LL USE IT:** *"I am like the textbook definition of a CONNECTED VIEWER right now with my laptop in front of me and my phone in my hand and the TV on right in front of me."*

**C**

### Coral World Park *(kor-uhl world pahrk)*, *noun*

Coral World Park is an über luxurious underwater resort located in Palawan, a small island chain in the Philippines. The resort's twenty-four suites will be located 60 feet below sea level and will include views of the ocean circle for 270 degrees. The suites, some will be for sale and can be built to suit the needs of the owner, will be referred to as "Anemones." This is a reference to the underwater animal sea anemones. One of the most unique things about Coral World Park, which is being developed by Singapore businessman Paul Moñozca and designed by ecoarchitect Jose Mañosa, is how guests will travel around. There will be hydro-powered mini submarines with glass bottoms that will transport guests around the resort. Coral World Park, which began construction in 2010 and is expected to open this year, will join the ranks of other underwater destination spots like Poseidon Undersea Resort in Fiji, the Al Mahara restaurant in Dubai, and Ithaa Underwater Restaurant in the Maldives. In addition to the underwater suites, Coral World Park will also have a hotel on land that will include a spa, casino, and even an underwater restaurant called Starfish.

**RELATED WORDS:** ecoarchitect, hydropower

**HOW YOU'LL USE IT:** *"Our dream honeymoon would be to spend a week in one of those underwater rooms at CORAL WORLD PARK in the Philippines."*

## craftermath *(KRAF-ter-math), noun*

The aftermath of an ambitious craft project. Thank the meteoric rise of photo-sharing social-media website Pinterest for popularizing this concept. After Pinterest debuted in 2010, it skyrocketed to become one of the biggest social-media sites on the market, growing more than 4,000 percent in 2012. The trendy image app and website has instilled in ambitious pinners an unyielding desire for crafting. Birthday cake in the shape of a dump truck? Wind chimes made out of bottle caps? You name it, they have to try it. Smitten with an unending stream of photos upon photos of DIY projects, users are testing out their skills in the real world and trying to recreate some of the things they've seen online. The result? A craftermath. It's that post-decoupage, post-cupcake decorating moment when all the glitter has settled and what you're left with are splinters of magazine clippings, smudged icing, smeared glue, and millions of shards of ribbon. And it's everywhere. *Glamour* magazine featured craftermath as its "Word of the Month" in its June 2012 issue, defining it as "the clutter left lying around after you finally make that thing you found on Pinterest." It wouldn't be surprising to see a craftermath storyline on Zooey Deschanel's *New Girl* or to see the word nestled among the scores of meta jokes at the online greeting-card website someecards.

**RELATED WORDS:** Pinterest

**HOW YOU'LL USE IT:** *"I tried to do some scrapbooking with the kids this morning but now it's a total CRAFTERMATH all over the dining room table."*

**C**

**cray** *(krey)*, *adjective*

Many have assumed that cray was merely a shortened version of the word crazy, since cray and cray cray (or kray kray) have been used in common speech the last few years. Since the release of the 2011 Kanye West and Jay-Z song, the song that popularized cray, another explanation of the word's meaning has emerged, which claims that the word isn't cray, but Kray. As the widely disseminated explanation goes, Kray was a reference to a set of twins with the last name Kray who committed a series of crimes in London in the twentieth century. The Kray twins are real, but the connection to the song by Kanye West and Jay-Z is probably fake. After the rumor about the lyric began to circulate, a girl named Mia Glenn stepped forward to say she had made it up. There had been some debate over cray versus Kray when the song was first released and in 2011 Glenn posted a fake explanation of the Kray connection on her Tumblr. Since then, the "explanation" has been reposted across the Internet, with many accepting Glenn's explanation as fact (and with nary a peep from the rappers). Whatever its real meaning, cray is most commonly used to mean crazy. Expect to hear more references to cray (and Kray) well into 2013 thanks to the song and the urban legends that now surround it.

**RELATED WORDS:** Kanye West, Kray, Jay-Z

**HOW YOU'LL USE IT:** *"You know what's CRAY? Our whole building lost power today so we all had to work in the dark until our laptop batteries wore out."*

**crowdfunding** *(KROUD-fuhnd-ing), noun*

Crowdfunding is a form of crowdsourcing, where instead of redistributing the work of a project to a large group of people, a request is made for a large group of people to donate funds to support a project. This is most frequently accomplished by setting up a website, so those interested can donate money online. The website Kickstarter is a popular example of crowdfunding at work, as the website allows users to create a page for their creative project as a way to earn the donations needed to fund it. The payments are processed through Amazon.com and, as such, both Amazon.com and Kickstarter take a small percentage of the money raised. Independent films, small businesses, political campaigns, and even band tours are all examples of things that can be supported by crowdfunding efforts. According to an online story by Kanyi Maqubela in the *Atlantic*, a report conducted by Crowdsourcing .org found that "more than 400 crowdfunding platforms were operating at the beginning of 2012, and several hundred more are expected to seek accreditation by the end of the year." Some have tied an increase in crowdfunding efforts with relaxed restrictions on investors under the United States of America JOBS Act, which became a law in the spring of 2012.

**RELATED WORDS:** crowd financing, hyper funding, JOBS Act, Kickstarter

**HOW YOU'LL USE IT:** *"My band never would have been able to go on tour if we didn't do some serious CROWDFUNDING on Kickstarter."*

**C**

**cyber-** *(SAHY-ber), prefix*

Pertaining to the Internet or computers, cyber is lifted from cybernetic and can be attached to any number of words to suggest an action happening online. There's cyberfeminism, cyberthief, and cyberpunk, or cyberespionage, which involves computer hacking as a way to gain access to information. A June 2012 piece by David E. Sanger in the *New York Times*, which surmised that the increase of covert attacks on other countries using sophisticated cyberattacks and cyberweapons from the country's cyberarsenal, uses eight separate cyber words. The people who launch these attacks are cyberinvaders; the preparation for battle is the cyberoffense. In Tehran, the military has created an elite branch, called the Cybercorps, to handle these attacks. "Cyberweapons, of course, have neither the precision of a drone nor the immediate, horrifying destructive power of the Bomb," Sanger wrote. "Most of the time, cyberwar seems cool and bloodless, computers attacking computers." The theory behind Sanger's piece is that such behavior could be sending us toward "mutually assured cyberdestruction" akin to nuclear war if we don't determine the specific criteria for their use. Looking forward, expect to see the development of more cyber words as we continue to navigate new developments utilizing the Internet.

**RELATED WORDS:** cyberespionage, Flame Computer Virus, -pocalypse, smishing

**HOW YOU'LL USE IT:** *"Mackenzie is just such a CYBERBULLY— she keeps posting mean things to my Facebook wall."*

## cyberespionage *(sahy-ber-ESS-pee-uh-nahzh)*, **noun**

Cyberespionage, which is sometimes written as two words or can be hyphenated as cyber-espionage, is the act of using spyware or malware to hack into another person's computer as a way to stealthily elicit information. The purpose of cyberespionage is to gain intelligence, particularly within a foreign government or company, however it has also been alleged that the Chinese government uses forms of cyberespionage to spy on its own citizens, especially those with ties to the United States. Cyber weapons used to carry out cyberespionage are things like the Flame worm, which when embedded in a computer can stealthily collect data and take screenshots of a person's computer. According to a story by Lucian Constantin on the website for PCWorld, Flame "is a very large attack toolkit with many individual modules. It can perform a variety of malicious actions, most of which are related to data theft and cyber espionage." In the summer of 2012, the Flame virus was found to be spying on computers in the Middle East. One fear about cyberespionage is that the predicted rise in cloud computing this year will leave our systems—and therefore our data—more vulnerable to spies and hackers. Currently, cyberespionage out of China and Russia poses the biggest threat to the United States.

**RELATED WORDS:** cloud computing, cyber Cold War, cyber spying, cyberweapon, Flame computer virus, phishing, smishing, Stuxnet

**HOW YOU'LL USE IT:** *"Forget things like bombs and missiles coming from other countries, now it's the destruction and fallout from CYBERESPIONAGE we have to worry about."*

**D**

**dark-matter detector** *(DAHRK mat-ter dih-TEK-ter), noun*

An über sensitive dark-matter detector located a mile underground in an old South Dakota gold mine will hopefully lead scientists to the discovery of dark matter. According to an Associated Press story by Amber Hunt in the summer of 2012, dark matter is "elusive matter that scientists believe makes up about 25 percent of the universe. They know it's there by its gravitational pull, but unlike regular matter and antimatter, it's so far undetectable." The name of the detector is the Large Underground Xenon detector—LUX for short—and its subterranean location is pivotal to its function, since normal laboratories make it difficult to detect something as sensitive as dark matter. The depth of the mine will help shield LUX from cosmic radiation that could interfere with its functions. The reaction to and coverage of news about the discovery of dark matter will most likely rival the news about the subatomic particle known as the Higgs Boson, which was announced in July of 2012. The dark-matter detector will no doubt be making headlines this year once it's turned on and starts to collect data.

**RELATED WORDS:** dark matter, Higgs Boson, LUX, WIMP

**HOW YOU'LL USE IT:** *"I heard that this could be the week that scientists will announce what they've found using the DARK-MATTER DETECTOR."*

**derecho** *(deh–REY-cho), noun*

D

A derecho is a very powerful and very large windstorm that occurs in a straight line over a long distance. In June of 2012, one such storm blew more than 600 miles from Chicago to Washington, D.C., causing widespread electricity loss and killing more than twenty. The word derecho means "straight" or "straight ahead" in Spanish; Gustavus Detlef Hinrichs is credited with first using the word in 1988 in the *American Meteorological Journal*. Derechos are sometimes described as being a "wall of wind," and usually include thunderstorms and downbursts. They can also be referred to as super derechos. Though the term isn't new per se, its usage has popped up in the American lexicon again as we continue to experience a surge of extreme weather conditions. This past year saw an increase in the number of record-breaking high-temperature days in the United States. Could 2013 see even more? In a 2012 story by the Associated Press, meteorologist Harold Brooks says that derechos and other "nontornadic wind events" are expected to increase as we continue to undergo climate change and endure the heat and instability that accompanies it. Some scientists are pointing to weather occurrences like this one as evidence of global warming.

**RELATED WORDS:** climate change, downbursts, global warming

**HOW YOU'LL USE IT:** *"I already have to climb into the bathtub every time a tornado passes through—now I have to worry about DERECHOS too?"*

**D**

**digital pills** *(DIJ-i-tl pilz), noun*

In what could very well mark a new era in medicine, the FDA has approved ingestible devices known as digital pills. Each pill contains a silicon chip that is about the size of a grain of sand. The chip contains magnesium and copper and can communicate with a patch that the patient is wearing on their skin. This sends messages to the patient's healthcare provider, who is thus monitoring their health from afar. Amy Maxmen writes in a post on the *Nature News Blog* that "digestible microchips embedded in drugs may soon tell doctors whether a patient is taking their medications as prescribed." If the person has forgotten to take their pill, then their doctor is going to know about it. As such, digital pills will be a way to help patients remember to take their medications. These digital pills have been developed by Proteus Digital Health and were tested using placebos in 2012. Expect to hear more about them this year when they make the leap from placebos to actual medications.

**RELATED WORDS:** Belviq, digestible microchips, pharmaceuticals, Proteus Digital health, smart pills, vitamins

**HOW YOU'LL USE IT:** *"I've been habitually pretty bad about remembering to take my medications so my doctor is switching me over to those DIGITAL PILLS, which he will be able to track."*

**Digital Public Library of America** *(DIJ-i-tl puhb-lik LAHY-brer-ee uhv uh-MEHR-i-kuh)*, **noun**

**D**

Launched in 2010 by Harvard University's Berkman Center for Internet & Society, the Digital Public Library of America is a nonprofit project that will bring together multiple sources and collections—like the Library of Congress and the Internet Archive—in one comprehensive digital platform. Some have expressed concern over the name of the project, saying it makes it seem like it's meant to replace the function of libraries. Organizers of the noncommercial project, which is also called the DPLA for short, say that it's meant to act as a resource for brick-and-mortar libraries—not as a replacement. As with Google Books, the DPLA will have to overcome copyright issues associated with digitizing works and could even require new laws from Congress. Librarian Robert Darnton of Harvard University is credited as the brains behind the project, which is expected to be up and running this year. Skeptics of the project point to things like an inability to define the project's scope and possible overlap with the efforts of Google Books and Project Gutenberg. Another concern is that since funds are being put into the DPLA, that means monetary support for ailing public libraries is being redirected.

**RELATED WORDS:** Google Books, libraries, Project Gutenberg, Scannebago

**HOW YOU'LL USE IT:** *"The DIGITAL PUBLIC LIBRARY OF AMERICA would really help me find sources for this book I'm working on."*

**D**

**double-dip recession** *(duhb-uhl-dip ri-SESH-uhn), noun*

If a recession is defined as the GDP shrinking for two consecutive quarters, then a double-dip recession is when it happens again after just a brief recovery period. After the U.S. economy struggled through the debt crisis, some economists and government officials feared that we hadn't seen the worst—that despite growth we were due for another recession. Another term for this economic pattern is a W-shaped recovery, since when plotted on a graph it looks like the letter W. This is also where the term double-dip comes from, since on the graph the line dips twice. Some economic experts are hesitant to use the term double-dip recession since it's not recognized by the National Bureau of Economic Research, the group that officially "dates" recessions. Instead, they assert that they are either two separate recessions or the continuation of the same one. Still the term persists in media coverage of the global financial crisis. "Federal Reserve has few options as economy flirts with 'double dip' recession" read a June 2012 headline for a story by Peter Morici on the Fox News website. Some point to fears of a taxmageddon in the United States, while others cite economic conditions in the UK and Europe as potential contributors for a double-dip recession to occur this year.

**RELATED WORDS:** debt crisis, Eurozone, taxmageddon

**HOW YOU'LL USE IT:** *"I don't know if my family can handle a DOUBLE-DIP RECESSION—we're already just scraping by as it is."*

## Ekso *(EHK-soh), noun*

**E**

Ekso is the name of an assisted walking system developed by the California-based Ekso Bionics company. Ekso uses a hydraulically powered exoskeleton system to help people walk. This particular model, which was first developed in 2010 and called eLEGS, is just one of many mobility systems being developed by Ekso Bionics. The original name for the system stands for "Exoskeleton Lower Extremity Gait System." A user is encapsulated in leg braces that include a hydraulic system that is operated by remote control. The product is already being tested at places like Mount Sinai and the company has said that it hopes to make Ekso available to consumers this year, at a cost of about $100,000. The Ekso has a battery life of about six hours and can help someone who has lost the use of their legs to walk, stand, and go from sitting to standing. In addition to helping people walk again, the exoskeleton technology is also being developed to assist soldiers and alleviate some of the burden of carrying a heavy pack load.

**RELATED WORDS:** Berkeley Bionics, Ekso Bionics, eLEGS, human exoskeleton

**HOW YOU'LL USE IT:** *"My cousin who's paralyzed from the waist down says he will definitely be buying the EKSO when it goes on the market, no matter what the price is."*

**E**

## El Bulli *(el–BOO–yee), noun*

The world famous Spanish restaurant, owned by Ferran Adria, may have closed its doors in 2011, but foodies everywhere will be talking about it again this year as it preps to take on a whole new life. Starting in 2014, El Bulli (which means bulldog in Catalan) will become the permanent home to the El Bulli Foundation, where new innovations in food and cooking are to be studied, concocted, and perfected; it has been referred to as a "culinary think tank." While the physical structure of the original restaurant remains, it will also be expanded, with new construction designed by the architect Enric Ruiz Geli. During its time as a restaurant, El Bulli (which can also be written as elBulli) was considered by many to be the greatest restaurant in the word, and was the recipient of a Michelin 3-star rating. The food was haute cuisine, served only for dinner and open just six months out of the year, and included pioneering techniques in molecular gastronomy. Despite its massive popularity, though, El Bulli never made a profit. When the center reopens in 2014, it will be eco-friendly, with the hopes of emitting no $CO_2$, and will allow for chefs, journalists, philosophers, and more to use the space as a way to learn more about the latest culinary techniques and to brainstorm new ones.

**RELATED WORDS:** molecular gastronomy

**HOW YOU'LL USE IT:** *"It's such a shame that EL BULLI closed, I always wanted to go there. I'm sort of hoping that after they open the new center in 2014 they'll start serving food again someday."*

**ermahgerd** *(erm-mah-gerd), interjection*

E

A purposeful mispronunciation of the interjection "Oh my god," erhmahgerd is used in a popular Internet meme that pairs silly photos (usually of animals or people making a surprised face) with the word ermahgerd in all capital letters and more similarly mispronounced language. The specific guidelines for writing the text in the meme has been likened to the phonetic spelling of someone speaking with a lisp or as though they have a retainer in their mouth. Other comparisons have been drawn to the speech style used by Kristen Wiig's popular *Saturday Night Live* character "Target Lady." Taking a popular example from the Internet, picture a photograph of a cat making a surprised face and looking at a box of Fancy Feast cat food. The caption above says "ERMAHGERD," while the words below read "FERNCER FERST," a mispronunciation of "Fancy Feast." Another photo showing a rabbit looking excitedly at carrots reads "ERMAHGERD KERRERTS." The meme is said to have originated with a photo of a slightly dweeby-looking girl with braces holding a trio of "Goosebumps" books, the caption reading "GERSBERMS" and "MAH FRAVRIT BERKS." (This translates to "Goosebumps" and "my favorite books.") Thus, this meme has also been referred to as "gersberms" or "berks," though ermahgerd has become the prevailing term. Erhmahgerd is used to show excitement and enthusiasm—both on Internet memes and in daily speech.

**RELATED WORDS:** berks, gersberms

**HOW YOU'LL USE IT:** *"ERMAHGERD you guys check out the awesome shoes I just got!"*

**E**

**EVEN Hotels** *(EE-vuhn hoh-telz)*, **noun**

As a way to meet customer demands for more health-conscious options while traveling, the InterContinental Hotels Group is launching a new chain of hotels this year called EVEN Hotels. The concept is that travelers on the go often fall off the fitness wagon while they're away. The rooms are designed to have things like fitness mats (and the space to actually use them) plus workout benches and a clothes rack that can double as a chin-up bar. According to a post on the Fodor's website by Stephanie Johnnidis, "The revolutionary concept is a far cry from your typical fancy-shmancy spa and wellness resort. Instead, these urban hotels will turn wellness travel mainstream by offering simple solutions to basic health needs at a price point comparable to a Courtyard or Four Points." The health-focused hotel chain, which is looking to open locations in cities like New York, Washington D.C., San Francisco, Los Angeles, and Boston, will include both the construction of new hotels and the purchase and conversion of old ones under the new brand. In addition to the in-room fitness amenities, there will also be things like healthier food options, hypoallergenic linens, antibacterial wipes, and LED dimmers. The rooms will also be designed to let in more natural light.

**RELATED WORDS:** ecoarchitect, ecotourism

**HOW YOU'LL USE IT:** *"I'm excited for these EVEN HOTELS that are opening, especially since every time I travel I feel like I gain 10 pounds because my whole workout routine gets messed up."*

**e-waste** *(EE-weyst), noun*

A shortened term for electronic waste, e-waste refers to discarded devices like computers, fax machines, printers, mobile phones, television sets, even refrigerators—anything electronic that doesn't stand a chance of decomposing in a landfill. In the past, it has also been referred to as Waste Electrical and Electronic Equipment (WEEE) but has since been recast as this friendlier, easier-to-remember term. One of the dangers of dumping e-waste into landfills is that it can leach contaminants such as lead, cadmium, beryllium, or brominated flame retardants. The state of Colorado passed a bill that will make it unlawful to dispose of e-waste in landfills, effective on July 1, joining states like Connecticut, Indiana, Vermont, and New Jersey (to name a few). "E-cycling" programs have now emerged as state agencies and health officials grapple with the increased need to handle a surplus of outdated technology. Combating e-waste has become a global issue, since much of the world's e-waste ends up in Africa. In March of 2012 at the UN Environment Programme headquarters in Kenya, a host of representatives from both the public and private sector met at the Pan-African Forum on E-Waste as the culmination of a three-year study and for a sort of call to action on combating this issue.

**RELATED WORDS:** e-cycling, e-scrap, Waste Electrical and Electronic Equipment

**HOW YOU'LL USE IT:** *"I'm trying to find a place where I can recycle all my E-WASTE—I have all these old computers just piling up in my basement."*

**E**

### Excalibur Almaz *(eks-KAL-uh-ber ahl-mahz)*, **noun**

British space company Excalibur Almaz has plans to launch a small group of civilian astronauts into orbit around the moon in the not too distant future. Sources cite the price tag as $150 to $155 million per person for the lunar fly-by and would include a decent amount of preflight training here on Earth. The mechanics of space travel aboard the Excalibur Almaz craft involve ion thrusters, "a high-tech propulsion system in which propellant is ejected using a solar field," according to a story in *The Economist* about the mission. The flight plan for the ship will take the passengers aboard a rocket to a space station, where they'll orbit the moon before reboarding a capsule and returning to Earth. In all, the space tourists will travel for at least six months since the planned propulsion system generates less thrust than a chemical-powered craft. The Excalibur Almaz joins a host of other space tourism ventures expected to launch in the coming years, including its United States–based competitor Space Adventures. The Space Adventures flight is in the same price bracket, though the length of travel is expected to be a little more than a week. Richard Branson's Virgin Galactic also has plans to launch humans into space this year, though for a much shorter period of time at just two-and-a-half hours.

**RELATED WORDS:** Google Lunar X-Prize, Soyuz, Space Adventures, space tourism, Virgin Galactic

**HOW YOU'LL USE IT:** *"I wonder which celebrities will shell out $150 million to travel past the moon aboard the EXCALIBUR ALMAZ spacecraft."*

**exergames** *(EK-ser-geymz), noun*

Meant to inspire sedentary kids (and adults) to get up off the couch, exergames are video games centered around virtual sports and exercise. The Wii Fit, which debuted in 2008, is a prime example of the fitness-driven exergame craze. Other options in the exergames category include Active Life: Extreme Challenge, Dance Dance Revolution, and EA Sports Active. Though one of the marketing claims for exergames such as these is to engage people in physical activity, a study published in *Pediatrics*, the journal for the American Academy of Pediatrics, has found that they're not as successful as many had hoped. "Exergames turn out to be much digital ado about nothing, at least as far as measurable health benefits for children," wrote Randall Stross in a June 2012 *New York Times* article. But the novelty and trend of exergames persists. Take, for example, the new Wii U and its game *Your Shape: Fitness Evolved 2013*, where users create a profile and workout plan that lets them set goals, work out, and even get recipes. Though health benefits of these games are still being debated, their use is on the rise: some predict that we might even start to see the use of exergames in schools in the coming years, not just as a way to encourage physical activity but as a teaching resource as well.

**RELATED WORDS:** childhood obesity, Wii U

**HOW YOU'LL USE IT:** *"I hate going to the gym and working out in front of other people so EXERGAMES have been really helpful to me."*

**F**

## Facebook depression *(FEYS-book dih-PRESH-uhn)*, *noun*

The social networking site Facebook allows millions of people to connect online by sharing photos and staying in contact despite large geographical distances. A side effect of all this information is becoming known as Facebook depression. Though not a clinical term, Facebook depression refers to the malaise and dissatisfaction felt after comparing oneself to the image put forth by others on Facebook. This could mean seeing a constant barrage of wedding and new-baby photos while you stew about being single, or reading updates from friends out having fun or traveling on vacation. What Facebook can't do is to put these events or updates in context—it's much harder to read a person's body language or to get the whole story from something they're posting online. Besides, when posting to a social networking site like Facebook, the inclination is to post things that put yourself in a positive light. But this skewed sense of reality is leading others to feel depressed that their lives aren't as great as their friends' lives are—or at least not as great as they look online. Discussions of Facebook depression are increasing in the wake of a string of teen suicides tied to alleged peer bullying via the site.

**RELATED WORDS:** FOMO, YOLO

**HOW YOU'LL USE IT:** *"Gotta love FACEBOOK DEPRESSION—just spent an hour looking at the vacation pictures of some girl I barely knew from college and now I'm totally jealous that my life isn't that great."*

## Facebook phone *(FEYS-book fohn)*, noun

There's been speculation about the development of a Facebook smartphone for years, and many are looking at 2013 as the year that the rumors finally come true. Facebook has reportedly hired engineers—many former Apple employees—specifically for their smartphone project, which used the codename "Buffy" and is being developed via a partnership with HTC. According to a blog post by Nick Bilton on the website for the *New York Times*, Facebook's reasoning for wanting to get in on the smartphone game would be pretty simple. "As a newly public company, it must find new sources of revenue, and it fears being left behind in mobile, one of the most promising areas for growth," Bilton wrote. Whether it's launched this year or not—CEO and founder Mark Zuckerberg has publicly denied its existence—expect to hear more about the Facebook phone. If it finally makes its big debut, it'll be one of the hottest, most buzzed about devices on the market. And if it doesn't come to fruition, the rumors will only become more fierce as pundits continue to try to predict its design, its concept, and ultimately, its release.

**RELATED WORDS:** Apple iTV, Facebook depression, FOMO, Project Glass

**HOW YOU'LL USE IT:** *"Facebook is a major reason why I've been able to stay in such close contact with family and friends while I'm living so far away, so I definitely plan to buy the FACEBOOK PHONE when it comes out."*

**F**

## fanboy/fangirl *(FAN-boi/FAN-gurl)*, *noun*

Any devoted follower of a particular product, brand, publication, and so on. Often used to reference avid readers of comic books or video game enthusiasts, it has now come to mean people who are obsessed with Apple products. Every electronic device they own is an Apple product, and they're quick to debate with anyone who will listen to why they think Apple products are superior to all others. "Attention all Mac fanboys/girls: Your chance to own a part of the Apple legacy is here," read the lead sentence to a June 2, 2012, story in *Time* magazine about a Sotheby's auction featuring six Apple-1 computers, which first went on sale in 1976. Not a derogatory term, instead it's a moniker claimed by the very community it denotes. "I'm such a fangirl," you may hear someone say as they maneuver an iPhone in one hand and an iPad in the other. The death of Steve Jobs in 2012 solidified for many a fanboy and fangirl that their devotion to the product is stronger than ever. Every year sees another round of rumors surrounding the latest Apple device or upgrade (the Apple TV set is supposedly due out this year), and with it, the fanboys and fangirls appear.

**RELATED WORDS:** fanatic, geek, nerd, superfan

**HOW YOU'LL USE IT:** *"You can always tell when a new edition of the iPhone is out because all the FANBOYS and FANGIRLS start to camp out in front of the Apple store."*

**farm to fork** *(fahrm too fawrk), noun*

Farm to fork is an extension of the locavore movement, where eating locally within the community is stressed. The defining factor of farm to fork is that the distance between where the food is grown and where it is consumed is as short as possible, skipping the middle man of shipping, processing, and distributing and instead going straight from the farm to the fork. Recent years have seen an increase in these practices thanks to the works of people like Michael Pollan, Wendell Berry, and further back to Alice Waters. The movement is closely linked to organic farming practices and community-supported agriculture, and is partially to thank for the uptick in things like urban farmers' markets. The farm-to-fork (or farm-to-table) concept has also inspired a number of restaurants centered around this ideal. The latest place to experience the farm-to-fork movement, though, is at weddings, as couples are electing to host their receptions in more of a rural setting, where the food that's served is grown just a few yards from where it's being consumed. As more eco-friendly and food-conscious couples wed this year, expect to hear more about the farm-to-fork movement as the latest trend touted by bridal magazines.

**RELATED WORDS:** agritourism, farm to fridge, farm to school, farm to table, locavore, vernacular food

**HOW YOU'LL USE IT:** *"We're thinking of having a FARM TO FORK wedding out in the Western part of the state because community-supported agriculture is really important to us."*

**F**

**fifth screen** *(fifth skreen)*, *noun*

The next big technology discussion for this year will be the idea of the fifth screen, and just exactly what it will be. You already have a TV (that's one) and a smartphone (two), plus a computer (three), and a tablet (four). What the fifth screen will be is the big question for 2013. Some have speculated that the fifth screen will be something along the lines of Google's augmented-reality eyeglasses, known as Project Glass. The futuristic-looking glasses place a tiny screen just above your field of vision, allowing you to interact with content the way you would on a phone. The device is expected to hit the market this year. Other technology experts are also predicting that a different kind of wearable screen will be your fifth screen in the coming years. Smart watches like the one coming soon from Pebble Technology or other wearable devices—also called wearables—are sure to be the next big thing. Fifth screen refers to the collective notion that we're not done developing these kinds of sophisticated devices and are due to experience more screen innovations in the coming years. A key element in the discussion of the fifth screen is that it creates another outlet for advertisers to connect with consumers, and would make ads that are integrated even more into our daily lives.

**RELATED WORDS:** connected viewer, Project Glass, wearables

**HOW YOU'LL USE IT:** *"When I was a kid the only screen we had was the one on the TV—now kids these days are onto a FIFTH SCREEN with surely more to come."*

**fiscal cliff** *(FIS-kuhl KLIF), noun*

F

A term coined by Federal Reserve chairman Ben Bernanke to describe fears of an economic breakdown in 2013 due to the simultaneous expiration of multiple tax-cut provisions. If we go off the fiscal cliff (i.e., Congress does not agree to extending these provisions, and/or cutting spending), economists predict a dire economic outlook for this year. Citizens will take a tax hike, many government services will either be scaled back or cut, and government spending in the private sector will be affected. "Are we really facing a fiscal cliff, a perfect storm, that is about to take us over the edge to financial disaster on a global scale?" asks Sheryl Nance-Nash in the opening sentence of her June 2012 piece in *Forbes*. Nance-Nash's description of the fiscal cliff as a perfect storm is a good way of putting it: A lack of investing begets a decrease in net worth which then leads to a drop in consumer spending. No one can catch a break, so the whole system starts to buckle. However, not all economists see this as a sure thing. Yes, the country is in a bit of a financial pickle—and has been for years—but detractors of this idea of a fiscal cliff assert that many are making it out to be a lot worse than it may turn out to be. A snappy name like fiscal cliff splashed across headlines certainly doesn't help either.

**RELATED WORDS:** Operation Twist, taxmageddon

**HOW YOU'LL USE IT:** *"All this talk about a FISCAL CLIFF made us nervous about our decision to buy a new house."*

**F**

## Flame *(fleym), noun*

Though it's been around for a few years, the 2012 discovery of the Flame computer virus across countries in the Middle East has thrust the term into the spotlight. A story by Ellen Nakashima, Greg Miller, and Julie Tate in the *Washington Post* called Flame "among the most sophisticated and subversive pieces of malware to be exposed to date." Here's how it works: the virus arrives on an affected computer either directly by way of a USB device or indirectly via phishing efforts, and can even show up under the disguise of routine software updates. Once implanted, Flame can access information on the affected computer and is advanced enough to log keystrokes, record Skype conversations, take screenshots, and access network traffic. Some have alleged that the cyberattack in the Middle East involving Flame, which primarily hit Iran, was done as a way to thwart nuclear weapon development there. Though the U.S. government hasn't confirmed it, some have alleged that the Flame attack on Iran was carried out by a partnership between the United States and Israel. Cyberespionage and cyberattacks like this one are a growing trend worldwide and the newest form of combat.

**RELATED WORDS:** cyberattack, cyberespionage, malware, Skywiper, Stuxnet

**HOW YOU'LL USE IT:** *"I feel conflicted about the FLAME computer virus because on the one hand, we're using it to get information about countries with nuclear weapons, but on the other hand, how long until someone uses it to spy on us?"*

## flash feast *(flash feest), noun*

The latest trend in spontaneous entertainment is sure to appeal to foodies and flash-mob fans alike. Originating out of Paris, a flash feast is a pop-up restaurant of sorts, where diners supply everything from the food and the silverware to the tables, chairs, linens, and more. The whole fun of a flash feast is that its location is keep a mystery until just before it starts, with participants converging on the area all at once to enjoy their meal, maybe do a little dancing, and then disappear as if it never happened. Partially to thank for the trend is the popular Diner en Blanc (dinner in white) in Paris, where all of the tablecloths, napkins, and even the participants' clothes are all white. Crowds can number in the thousands at the annual Parisian event, which began in 1988. The flash-feast trend, which is also referred to as a mob banquet or a meal mob, is catching on across Europe, Asia, and Australia, and in U.S. cities like New York, Boston, Chicago, San Francisco, and more. Expect to hear more about them as the trend grows.

RELATED WORDS: cash mob, diner en blanc, flash mob, meal mob, mob banquet, night market, pop-up picnic

HOW YOU'LL USE IT: *"We are planning our trip to Paris to coincide with that famous FLASH FEAST they do every year, since it's always been a dream of mine to attend."*

**F**

### floordrobe *(FLOHR-drohb), noun*

At the end of the work day when you finally make it home and change into your pajamas, don't bother hanging up your clothes in the closet. Simply drop them on the floor with everything else you've worn that week, into the growing pile of clothes known as the floordrobe. A common occurrence among teenagers and college students—as every parent can attest—a floordrobe is a pile of clothes, shoes, belts, and more that collects on the floor of a person's bedroom over time. Perhaps the door can no longer open all the way, or a path from the door to the bed has been cleared. A floordrobe hashtag is getting trendy on the popular photo-sharing site Instagram, where people post pictures of their own clothing heaps. Floordobe is a combination of the word floor with wardrobe, and was used as early as 1994 in a story by Bob Levey in the *Washington Post*. These days, with more college grads returning home after graduation and taking their messy habits along with them, the floordrobe carries on.

**RELATED WORDS:** Generation Debt

**HOW YOU'LL USE IT:** *"Mom, my room is not messy—I've just been cultivating a FLOORDROBE for a few weeks."*

**FOMO** *(FOH-moh), acronymn*

The fear of missing out, as a social concept, is not a new idea, but with the massive increase in social-media sites, the concept has become a legitimate point of discussion among psychologists and those observing the effects of the Internet on social behavior. FOMO is the acronym for the "fear of missing out," and it was the subject of a 2011 story in the *New York Times*. FOMO occurs when a person feels the need to check Facebook, Twitter, Four Square, Instagram, and other social-media websites to see what others are doing and to check the feedback (how many likes, how many comments) of something they've posted. Often, the FOMO phenomenon can happen while a person is already out in a social setting. Despite where they are or what they're doing, there's this urge to see what other people are up to. How many times have you seen a table of people at a restaurant or a crowd of people at a bar, each with their face glued to their smartphone? That's FOMO, and it can be addicting. With new social media apps launching all the time, the FOMO affliction is sure to continue. It wouldn't be surprising to see legitimate psychological studies exploring the effects of FOMO in the coming years or even a backlash among users who battle FOMO by forcing their friends to check their smartphones at the door.

**RELATED WORDS:** Facebook depression, YOLO

**HOW YOU'LL USE IT:** *"I really hate it when you're out with friends and FOMO totally takes over—everyone's sitting around not talking to each other with their eyes locked on their phones."*

**foraged** *(FAWR-ijd), adjective*

Foraged is likely to be a hot term this year, showing up on menus in restaurants emphasizing locally grown, sustainable, and farm-to-table philosophies. The standard meaning of forage is to look for food, and as such, menus that say certain ingredients have been foraged will mean that the waitstaff, or more likely the chef, will have harvested these items themselves from somewhere nearby. You may also see this referred to as "chef dug." The idea of a foraged meal not only lends itself to the locavore movement (since the chef can't go very far) but it also gives the diner a sense of special treatment and exclusivity. Furthermore, foraged foods play into the notion that there is more that is edible out in the wild than we may realize. Foraged foods may include things like wild greens, berries, and mushrooms, or even seaweed, dandelions, ramps, and more. If it's wild and it's edible, then it can be foraged.

**RELATED WORDS:** chef dug, farm to fork, root to stem

**HOW YOU'LL USE IT:** *"Our wedding will feature a salad for the first course that has been FORAGED by the chef from nearby meadows and woodlands."*

**fracking** *(FRAK-ing), noun*

F

Also known as hydraulic fracturing, this method of stimulating subterranean reservoirs of oil or natural gas involves pumping a large amount of water, sand, and chemicals deep into the ground at a high pressure to produce a vertical fracture. Fracking can happen anywhere from 5,000 to 20,000 feet below the surface, releasing the oil or gas for extraction. It is a controversial practice due in large part to the risk of contaminating ground water and negatively affecting air quality. There are also concerns that those working on the sites are at risk due to exposure to dust containing respirable crystalline silica. Companies that partake in fracking maintain that the practice is a safe one. The 2012 documentary *FrackNation* by filmmakers Phelim McAleer and Ann McElhinney explores this practice, mostly in a positive light, while Josh Fox's 2010, Academy Award–winning documentary *Gasland* took a look at the opposition. Expect to hear more about fracking after the Department of the Interior determines regulations on how drilling companies should seek approval for sites in the United States, where it has estimated more than 2,500 trillion cubic feet of natural gas waits underground. Aside from fuels, fracking can also be used to access groundwater and as a method for mining, among other things.

**RELATED WORDS:** hydraulic fracturing, hydrofracking, induced hydraulic fracturing

**HOW YOU'LL USE IT:** *"I've seen both* FrackNation *and* Gasland *and I still don't know where I stand on FRACKING."*

**F**

**fuel cell** *(FYOO-uhl sel), noun*

Apple's North Carolina data center, which is expected to be finished by the end of the year, will include the biggest nonutility fuel cell facility in the country. A fuel cell is an energy-creation device that converts biogases into electricity, the byproducts of which are simply heat and water. There has long been talk of creating cars that can run on fuel cells—President George W. Bush even created a program back in 2003 meant to help develop this technology. Expect to hear more about this green energy source as Apple nears completion of its building project. In addition to the fuel-cell facility, across the street from the 500,000-square-foot building will be a 100-acre solar farm, which in conjunction with the fuel cells will allow the building to run completely devoid of coal power. This comes at a key time for Apple, who in the past has been criticized by the environmental nonprofit Greenpeace for being too dependent on coal. The facility in North Carolina is one of three eco-friendly building initiatives by Apple, with data centers in California and Oregon running on clean energy as well. Expect to hear more about fuel cells and other alternative energy sources as more companies and building projects follow suit.

**RELATED WORDS:** biogases, clean energy, GreenGT, solar farm

**HOW YOU'LL USE IT:** *"I'd like to see all major companies transition away from coal power and into the use of FUEL CELLS, solar power, and wind turbines as a way to speed up the green energy process."*

# G
# H
# I

**G**

## GAIA mission *(GAHY-yuh mish-uhn), noun*

This ambitious project from the European Space Agency set to launch sometime during the summer—many sources say August—will attempt to create a 3-D map of the Milky Way galaxy. Think of GAIA, which is actually an acronym for Global Astrometric Interferometer for Astrophysics, as a galactic census to occur over a five-year period. (Gaia was also the name given to the goddess of the Earth in ancient Greece.) When the craft is launched, it will use two high-powered telescopes and a billion-pixel camera while it rotates and scans, in the process documenting the positions, movements, and distances of a billion stars, as well as discovering new stars, planets, and asteroids in the process. A billion stars may sound like a lot, but it's actually only 1 percent of all the stars in our galaxy. Expect to hear about GAIA in the news when the mission is launched and also in years to come as the data about its findings begin to pour in. "Many previous space missions would hang onto the data and then send it in short bursts. However, GAIA can't afford to cache the amount of data it produces so will have to transmit constantly," said Jon Kemp, marketing and applications manager for E2V, to *Wired* magazine. E2V is the UK company that created the billion-pixel camera to be used on the mission.

**RELATED WORDS:** astrometry, E2V, space archaeology

**HOW YOU'LL USE IT:** *"Have you seen the latest photos from the GAIA MISSION—it found five new asteroids!"*

**-geddon** *(GED-n), suffix*

G

Taken from Armageddon, the end of days described in the Bible as the battle between good and evil. Tacking this suffix onto just about any word refers to a seemingly decisive situation rife with conflict. The much-hyped taxmageddon, for one, refers to the potential doomsday scenario following expected tax hikes this year. Similarly, Eurogeddon describes the debt crisis that has overcome much of Europe. The -geddon suffix is adaptable to a host of situations wherein the worst case scenario outcome looks pretty grim for all involved. When news hit that Yahoo.com was debating massive layoffs of their employees, that become Yahoo-mageddon. When California restaurants were banned from serving foie gras in 2012, that was foie-mageddon. After a photo showing J. Crew President and Creative Director Jenna Lyons painting the toenails of her young son was featured on the company's website, the controversy that ensued was dubbed "toemageddon." The word was most famously used by Jon Stewart, the host of the satirical news program *The Daily Show*, who lambasted the media for making a big deal out of a seemingly non-newsworthy story. A similar word-creation practice exists when coining new terms ending in "-pocalypse," which is lifted from the word apocalypse. This common practice—used extensively by the media—can perhaps be most famously traced back to the use of the suffix "-gate" (from the word Watergate)—as in spygate and nipplegate—which is used when referring to a scandal.

**RELATED WORDS:** -pocalypse, taxmageddon

**HOW YOU'LL USE IT:** *"It is about to be a lunchmaGEDDON over here if Steve doesn't come back with the burritos."*

**G**

**gendercide** *(JEN-der-sahyd)*, *verb*

Gendercide, which is also known as sex-selective abortion, is the term used to describe the early termination of a pregnancy based on the gender of a fetus. The term is most often used in situations when a female baby is aborted, as is often the case in countries like China or India where cultural pressures favor male children. In May of 2012, the House of Representatives rejected a bill—called the Prenatal Nondiscrimination Act—that would have banned sex-selective abortions in the United States and would have imposed fines and prison sentences on doctors who assisted in these services. Elsewhere in the world, countries like Canada, the United Kingdom, France, Germany, and Switzerland already ban the practice. Abortion and reproductive rights continue to be a hot-button topic, especially since it is an issue that is so polarizing. Women's reproductive rights are a major talking point among U.S. politicians, especially during the time following an election year when candidates include their stance on the issue as part of their platform. As such, the conversation about which limitations—if any—can be imposed on women's reproductive rights by federal mandate will continue into the year.

**RELATED WORDS:** Prenatal Nondiscrimination Act, sex-selective abortion

**HOW YOU'LL USE IT:** *"Making sure your views on GENDERCIDE sync up with that of your spouse can be pretty important down the road."*

**Generation Debt** *(jen–uh–ray–shuhn DET)*, **noun**

Economic troubles percolating through the last few years have helped to brand those born between 1980 and 1994 as Generation Debt. The term can be largely traced back to Anya Kamenetz's 2006 book *Generation Debt: Why Now Is a Terrible Time to Be Young*. Since then, the idea has been picking up steam as members of this group struggle to find jobs, pay their student-loan bills, and do things like buy property or a new car. Many have had to move back home with their parents as a way to save money, and consistently shop around for the best price before making a purchase. Also referred to as the Millennial generation or the broke generation, this segment of the population has come of age during a particularly trying time, when high interest rates on student loans and high unemployment rates have combined to create a sort of perfect storm. It's the first time that a generation is poised for less financial success than its parents. Expect to hear more talk about Generation Debt this year as politicians debate government policies concerning student loan interest rates, financial pundits advise how to get out of the debt cycle, and as more studies about this cultural enclave continue to surface.

**RELATED WORDS:** Generation Broke, Generation Screwed, Millennial generation

**HOW YOU'LL USE IT:** *"As members of GENERATION DEBT, we may not be able to pay our bills or get decent jobs, but at least our age group has a snappy new name."*

## GENI *(JEE-nee), noun*

**G**

GENI is an acronym for the Global Environment for Network Innovations, a $40 million über-fast Internet built by the National Science Foundation that connects next-generation broadband networks. GENI is meant to serve as a sort of "sandbox" for innovators and developers taking part in the U.S. Ignite initiative, a project aimed at the innovation and creation of new high-speed Internet applications. The introduction of this lightning fast network came about in 2012 when President Obama signed an executive order to make broadband construction cheaper and more efficient by allowing for new cable to be laid during highway construction. The National Science Foundation calls GENI "a unique virtual laboratory" that is meant to allow for experimentation and innovation, leading to advancements in education, healthcare, energy, public safety, and more. Allowing developers to have access to this resource on such a large scale means that projects can be evaluated and deployed earlier, and that their socioeconomic impact can be better determined. Access to the network is expected to become available over the course of the next five to six years in more than twenty-five cities.

**RELATED WORDS:** Google Fiber Project, National Science Foundation, U.S. Ignite

**HOW YOU'LL USE IT:** *"I wish I had gone to school to be a web developer or some sort of Internet guru so I could be part of this GENI project."*

**geoengineering** *(jee-oh-en-juh-NEER-ing), noun*

**G**

Geoengineering is a field of study that seeks to thwart, inhibit, and learn about climate change via manmade intervention efforts. Harvard University geoengineers David Keith and James Anderson will attempt to mimic the cooling effects of a volcano eruption on the planet this year by releasing sulfate aerosols into the atmosphere. Thousands of tons of this sun-reflecting chemical will be released via balloon some 80,000 feet above New Mexico. Other geoengineering strategies include injecting water and other chemicals that would help deflect the sun's rays into the atmosphere. The idea is that doing this will reduce greenhouse gases and thus cool the planet. Other geoengineering techniques involve encouraging phytoplankton blooms in the ocean as a way to keep carbon in the seabed. Opponents of geoengineering say that interfering could produce more harm than good, disrupting the planet's natural hydrologic cycle, inhibiting rainfall, and turning the sky white. The concept of geoingineering has been around for years, as a combination of geoscience and engineering. It has also been referred to as climate engineering, climate remediation, and climate intervention. As large-scale projects start to happen and receive media coverage in the coming years, expect to hear more about this practice in a big way.

**RELATED WORDS:** climate engineering, climate intervention, climate remediation, global warming, greenhouse effect

**HOW YOU'LL USE IT:** *"After my little sister saw Al Gore's movie* An Inconvenient Truth, *she decided to study GEOENGINEERING and solve climate change."*

**G**

**gesture-control technology** *(JES-cher kuhn-trohl tek-NOL-uh-jee), noun*

One of the biggest technological advancements that everyone will be talking about this year is gesture-control technology, which allows you to control something without touching it. The use of sensors allows devices with gesture-control technology to be navigated by simply moving your hand or body. Gesture-control technologies are already being applied to things like Flutter, a Mac app that can control your music by waving your hand in front of the webcam. There's also the Xbox 360 with Kinect, in which your body's motions are what control the game on screen. Samsung's Smart TV can also be controlled by a mere wave of the hand. You can expect to see the use of this technology more this year, and in a myriad of places. For one, there's Leap Motion, created by Michael Buckwald and David Holz. Leap Motion has developed a more sophisticated application of gesture control technology that can be used on its own in place of a mouse, or applied to preexisting devices by developers (things like smartphones, laptops, tablets, and the like). It is expected that Leap will be adapted to a wide range of applications in 2013.

**RELATED WORDS:** exergames

**HOW YOU'LL USE IT:** *"I should probably get a TV with GESTURE CONTROL TECHNOLOGY since I lose my remote at least once a week."*

## Godello *(goh-DEY-oh), noun*

**G**

Godello is a type of Spanish white wine that is poised to make a comeback this year as its popularity grows among oenophiles, sommeliers, and novice wine lovers alike. An article by food and wine writer Don Rockwell in the the *Washingtonian* described Godello wines in 2007 thusly: "A great Godello combines the minerality of a great Chablis with the acidic snap of a Sauvignon Blanc—it comes at you quietly, with elegance and persistence." Godello, which is from the autonomous Galicia region of Spain, is said to pair well with shellfish. The wine decreased in popularity around the 1970s when the grape became overshadowed by the Palomino. But within the last few years, it's begun slowly coming back. Publications like the *Financial Times*, the *Huffington Post*, and the *New York Times* all included coverage of the Godello as a wine worth trying during the summer of 2012, inciting talk that this bright white is due to be the next big thing in the wine world. Wine writer Jancis Robinson of the *Financial Times* wrote of the varietal, "Forget Albarino. Godello is the hot new north-west Spanish grape variety with great class." Godello, which is also the name for the grape used to the make the wine, can be called a host of other names including Gouvei, Ojo de Gallo, Trincadente, and Verdello.

**RELATED WORDS:** oenophiles, wine

**HOW YOU'LL USE IT:** *"Let's try that new Spanish restaurant up the street—I hear they have a great GODELLO on the wine list."*

**G**

### Google Fiber *(goo-guhl FAHY-ber), noun*

Google Fiber is an Internet and cable service that features superfast connection speeds, loads of storage, and crisp HD. According to the Google Fiber website, the service is reportedly 100 times faster than the average broadband Internet we're used to, working at 1,000 Mbits per second. As for the TV portion of the device, the controller is a Google Nexus 7 tablet (which can also be used on its own). The cable service includes HD channels and on-demand shows and movies, plus access to Netflix and YouTube straight from the TV. The device's storage is also quite large, allowing you to record up to eight programs at once and store nearly 500 hours of HD video on its 2TB hard drive. Google Fiber made a splash in 2012 when it opened up registration to various communities in Kansas City. These neighborhoods, or "fiberhoods" as Google is calling them, will start to get service sometime this year if the project is able to get the kind of response it needs. If all goes well, service to the rest of the country could follow not too long after.

**RELATED WORDS:** fiberhood, GENI, gigabit, undergrounding, U.S. Ignite

**HOW YOU'LL USE IT:** *"I am so excited for GOOGLE FIBER, I plan to be the first to sign up once the service comes to my city."*

**Google Lunar X-Prize** (*goo-guhl loo-ner EKS prahyz*), *noun*

**G**

The Google Lunar X-Prize is an international space race of more than two-dozen teams competing to put a robot on the moon and win part of a $30 million prize cache. Each team's lunar rover is expected to traverse 500 meters of the moon, taking video, images, and other data and transmitting it all back to Earth. Per guidelines from NASA, the lunar rovers are to preserve past moon-landing sites so as not to destroy human boot prints or rover tracks from the Apollo missions. One of the bonus prizes being offered actually involves photographing these historic sites. This concept of incentive prizes for private development has been on the rise in recent years as a way to engage a new cadre of trailblazers who wouldn't necessarily fit within the typical grant model. Though the project first launched in 2007, coverage of the Google Lunar X-Prize is ramping up as the teams draw closer to the end date, which was extended from December 31, 2012, to December 31, 2015. Talk of the competition is also increasing since China's Chang'e 3 mission plans to put a lunar rover on the moon this year. Some are speculating that China's mission will speed up efforts of the teams vying for the Google Lunar X-Prize.

**RELATED WORDS:** Ansari X-Prize, Chang'e 3 mission, Lunar Lander Challenge, X-Prize Foundation

**HOW YOU'LL USE IT:** *"I wonder if one of the teams competing for the GOOGLE LUNAR X-PRIZE will put a robot on the moon before China's Chang'e 3 mission does."*

**G**

**graphene** *(GRAF-feen), noun*

The super-strong material graphene, which consists of just one layer of honeycomb-structured carbon atoms, started to get a flurry of media coverage in 2011 after a team studying graphene was awarded the Nobel Prize in 2010. Now, talk of graphene is increasing as companies like Focus Metals claim plans to produce graphene in commercial applications this year. Graphene has been called a "miracle material" by some due to its strength and its ability to heal itself. It is the strongest material ever (even beating out the diamond), is great at conducting electricity, and even has self-cooling properties. Potential applications for graphene include solar cells, touchscreen devices, cameras, transistors, and computers—some have claimed that it could even replace silicon in electronics. It does have its drawbacks, though. Rebecca Boyle writes in *Popular Science* that "it can't be arranged simply in a lab—it must be grown just so—and it's highly reactive with other compounds, including itself." For all the wonders of grapheme, though, it may seem like a space-age material encountered solely by scientists in labcoats. But in fact, we've probably all encountered graphene in a much more common form: multiple layers of graphene sheets combine to form graphite, found most commonly in pencils.

**RELATED WORDS:** graphite, Nobel Prize, silicon, Silicon Valley

**HOW YOU'LL USE IT:** *"If the commercial production of GRAPHENE surpasses silicon, do you think they'll rename Silicon Valley?"*

**gravity tractor** *(GRAV-i-tee trak-ter)*, *noun*

G

As more programs like the B612 foundation ramp up projects and technologies aimed at identifying near-Earth asteroids, so too does the discussion about how to handle them. One such concept on how to deter an asteroid bound for Earth is known as a gravity tractor. The idea behind a gravity tractor is to knock an asteroid off its course simply by hovering over it, using the gravitational attraction of the craft to alter the asteroid's path. Such a plan was developed by a pair of NASA astronauts, named Edward Lu and Stanley Love, and is still in its conceptual phase. But as things like the GAIA mission and the B612 Sentinel get off the ground and begin to discover new space rocks, there's no doubt that the conversation will then turn to how we should manage them. Other theories involve using nuclear weapons to blow up asteroids, landing crafts on them *Armageddon*-style and then pushing them off course with the force produced by a spaceship's engines, or even simply crashing a spacecraft into the asteroid in the hopes of breaking it up. A gravity tractor eliminates a lot of the risks associated with these other ideas, especially because it doesn't need to make contact with the asteroid in order to be successful.

**RELATED WORDS:** B612 Sentinel, GAIA mission, Sentinel Mission

**HOW YOU'LL USE IT:** *"A GRAVITY TRACTOR may sound like something out of a science-fiction movie but it could possibly be the thing that saves us from being destroyed by an asteroid."*

**G**

**Great Pacific Garbage Patch** *(greyt puh–sif–ik GAHR–bij pach),*
   *noun*

The Great Pacific Garbage Patch is a massive collection of trash
that has amassed in a remote part of the Pacific Ocean. Located
about 1,000 miles to the northeast of Hawaii, the patch is by some
estimates about twice the size of Texas. A garbage patch like this
isn't the only one of its kind, either, as oceanographers estimate
that there may be five of these located all over the world at various
gyres—places in the ocean where heavy currents and winds meet to
form a kind of swirling whirlpool. The majority of the trash is plastic that has washed out to sea, some of it breaking down into rice-
sized bits called microplastic. This poses a big risk to fish, who eat
the chemical-laden microplastic and can't digest it, and for people
who eat the fish and all the toxins along with it. Project Kaisei is
an initiative by the Ocean Voyages Institute that aims to study the
Pacific Garbage Patch and clean it up—perhaps even turning the
waste into diesel fuel. The Great Pacific Garbage Patch, which is
also called the Pacific Trash Vortex, was discovered in 1997 by
Charles Moore, though its existence had been predicted as early as
1988. This will become a term to know this year as debris from the
2011 Japan tsunami is expected to join it in the near future.

**RELATED WORDS:** gyre, Marine Drone, microplastic, Pacific
Trash Vortex, pollution, Project Kaisei

**HOW YOU'LL USE IT:** *"My little brother decided he wanted to
go to school to study environmental biology and oceanography
after seeing a segment on the news about the GREAT
PACIFIC GARBAGE PATCH."*

**GreenGT H2** *(GREEN jee tee eych too)*, *noun*

G

The GreenGT H2 is an electric- and hydrogen-powered racecar built by high-concept automaker GreenGT. According to a story in Auto World News, the GreenGT H2 "is powered by a hydrogen fuel cell, which produces the electricity to drive two electric motors pushing out the equivalent of 540 hp." For gearheads feeling guilty about burning through gasoline, the GreenGT H2 is certainly a vehicle worth drooling over. Its high-power fuel cell can supply the car with enough energy to hit speeds close to 185 mph. And since the car runs on hydrogen—which, it should be noted, is less flammable than gasoline—it means the car emits no pollutants or other harmful emissions. GreenGT is a Franco-Swiss car company whose main objective is to build vehicles that run 100 percent clean with hydrogen. The GreenGT H2 will test its limits this year when it takes part in the 24 Hours of Le Mans, a more than 3,000-mile-long race circuit in France that takes place over the course of twenty-four hours. The GreenGT H2 will serve as a model for biofuel as an energy source during this endurance race, which is held yearly in June.

**RELATED WORDS:** fuel cell

**HOW YOU'LL USE IT:** *"I'm planning to take a trip to France over the summer to catch the GREENGT H2 racing in the 24 Hours of Le Mans."*

**G**

### Grexit *(GREK-sit), noun*

In simple terms, the Grexit is the Greek exit from the Eurozone, a word that was developed by Citigroup's Ebrahim Rahbari and Citi Chief Economist Willem Buiter in 2012. Since the word's inception, the usage has spread. Throughout 2012, economists speculated on the likelihood of the Grexit by 2013. Greece's debt is due in part to borrowing large sums of money from what's referred to as the troika, a trio of foreign creditors that is comprised of the European Commission, the European Central Bank, and the International Monetary Fund. The country is also afflicted by a high unemployment rate of 22 percent and has been forced to make widespread cuts in the public sector as a way to close the debt gap and make good on promises to right its economic woes. According to a CNN article by Teo Kermeliotis titled "'Grexit' Worries Fuel Nation's Vicious Circle," a "return to the drachma, Greece's currency before the euro, could drastically cut the value of existing cash." Many Greeks have been pulling their money out of the banks for fear of the drachma's return, which could have the effect of causing the very thing that they fear. Whether the Grexit has happened already or is still a possibility for debt-riddled Greece, it's a word we'll be talking about well through this year as experts surmise the continued side effects throughout Europe.

**RELATED WORDS:** austerity, bailout, drachma, Eurozone, troika

**HOW YOU'LL USE IT:** *"I sort of think that after the GREXIT would be the prime time to take a trip to Greece since everything will probably be so cheap there."*

**grey babies** *(grey BEY-beez), noun*

G

Grey babies, or shades of grey babies, refers to the potential baby boom that could happen this year following the mainstream popularity of E. L. James's erotica book series *Fifty Shades of Grey*. Originally released in 2011, the book series became a bit of a cultural phenomenon once the final book, *Fifty Shades Freed*, hit shelves in 2012. Countless newspapers across the country reported on this "mommy porn" trend, since the majority of its audience seemed to be suburban housewives. Even *Saturday Night Live* spoofed it with a Mother's Day sketch about middle-American moms getting caught reading the book. It wasn't long before talk of the book turned to what the sexually explicit read would lead to: babies. But not everyone thinks that this baby boom is imminent. *Slate* writer Amanda Marcotte counts herself as a skeptic of this idea, saying that while she's sure grey babies will occur, the phenomenon won't be that widespread. "The suggestion that a popular erotic novel is enough to get so many engines turning is also a tad hard to believe," Marcotte says. "The porn industry and romance novel industry have both been enormous for decades now." Either way, the concept is one of those pop-culture trend stories that will be sure to pop up in the media this year.

**RELATED:** cliterature, mommy porn

**HOW YOU'LL USE IT:** *"Three women in our neighborhood are all due around the same time next year—their GREY BABIES were born about nine months after our book club read* Fifty Shades of Grey.*"*

**G**

**grey divorce** *(grey dih-VOHRS)*, **noun**

A growing trend among couples over the age of fifty, a grey divorce refers to the dissolution of a long-term marriage when the couple is older, and thus have grey hair. It has been cited lately as a trend among baby boomers who, after enduring the hustle of childrearing, are finding that they no longer relate to one another. The divorce rate in the United States has grown to one in four for couples over the age of fifty. Some experts have cited multiple factors that can enable couples to go their separate ways, such as a longer life expectancy, the social acceptability of divorce, infidelity, and greater financial independence among women, who are more often becoming the initiators of the divorce. A commonly cited example of grey divorce is when former Vice President Al Gore and his wife Tipper separated in 2012 after forty years of marriage. Arnold Schwarzenegger's separation from his wife of twenty-five years, Maria Shriver, is another example. A recent movie starring Meryl Streep and Tommy Lee Jones explores the idea of a couple who seems to be headed for a grey divorce. Called *Hope Springs*, the movie shows the difficult challenges facing a middle-aged couple who decide to attend a weeklong marriage counseling seminar. Grey divorce isn't just a phenomenon that is happening in America, either. In Japan, this trend is often referred to as "retired husband syndrome."

**RELATED WORDS:** baby boomers, retired husband syndrome

**HOW YOU'LL USE IT:** *"Did you hear about Michael and Kelly up the street? Thirty years together and now they're going through a GREY DIVORCE."*

**growth hacker** *(GROHTH hak-er), noun*

As the needs of startups, Internet companies, and other digital platforms change, it is the natural ebb and flow for some positions to become obsolete while others are created. One new emerging trend on the tech scene of Silicon Valley is the idea of a growth hacker. Someone who holds this position straddles both the marketing department and the programming side of the company. This is a person who understands how to market a product, but who also has the technical knowledge of a programmer and can do things like run A/B testing and evaluate code. At its core, a growth hacker is simply someone intent on strategizing ways to grow a product and a company, and who has the technical skills needed to work with developers and make the appropriate changes or improvements to deliver what their audience wants. The term growth hacker was developed by Sean Ellis, the founder and CEO of CatchFree, who previously spent time working at places like Dropbox and Eventbrite. The concept of a growth hacker started to pick up steam thanks to a blog post by Andrew Chen, an entrepreneur and blogger who writes about the Internet and has experience in digital media, advertising, and various startup companies. Expect to hear more about these positions as companies get on board with the trend and post job openings for their own growth hackers.

**RELATED WORDS:** digital media, hacker, Internet, new media, Silicon Valley

**HOW YOU'LL USE IT:** *"My boss just put up a job posting for a GROWTH HACKER after he read an article online saying that more and more tech companies have them."*

**G**

**gTLDs** *(jee-tee-el-DEEZ)*, **noun**

Internet top-level domain names like .com, .biz, .net, and .org will be getting some interesting company this year following the sale of gTLDs, which is short for generic top level domains. Sometimes referred to colloquially as dot-words, these new website addresses could end in just about anything if the Internet Corporation for Assigned Names and Numbers (ICANN) approves the application. Acquiring a gTLD of one's own doesn't come cheap, though—the application fee alone is $185,000 and the price to renew annually is $25,000. Among those heavy hitters to apply for gTLDs with ICANN was Amazon, who logged requests for things like .App, .Cloud, .Free, .Kindle, and .Search, spending more than $14 million in the process. Amazon.com has some competition for acquiring the .App gTLD, since ICANN logged thirteen separate applications for its ownership. Microsoft filed requests for gTLDs like .Hotmail and .Skype while L'Oreal applied for .Hair, .Makeup, and .Beauty. Apple logged just one application with ICANN: .Apple. According to a story by Chris Barth in *Forbes*, "not all applied-for gTLDs are so formal, though. Three separate organizations applied to own the .Sucks domain, for example, perhaps envisioning a world where people go to www.microsoft.sucks to voice their complaints." After ICANN reviews the applications and makes their decisions, we can expect to start putting these new gTLDs to use in the spring of this year.

**RELATED WORDS:** dot brand, dotcom, dot-words

**HOW YOU'LL USE IT:** *"It's kind of confusing trying to remember all these new GTLDs instead of just typing in .com like we used to."*

**guayule** *(why-YOO-lee), noun*

G

Expect to hear the word guayule more in the news as the constant search for alternative, renewable resources continues. Guayule is the colloquial term given to Parthenium argentatum, a flowering desert shrub found in Mexico and the southwestern United States. A byproduct of guayule is natural latex, which also happens to be hypoallergenic, and can be used to make things like rubber gloves, balloons, and more. In fact, the name guayule comes from an Aztec word that means "rubber." Some have also referred to guayle as "the fuel of the future" because the stems and branches from the plant can be used as a source of energy—even after the natural latex properties have been extracted. Advocates for the use of guayule as a fuel say that since it's not used for food, there's no competition for its use. Another added benefit? Since it grows in dry climates, it thrives where other fuel sources and crops cannot. In the summer of 2012, the U.S. Department of Energy granted $5.7 million to Ohio State University's Ohio Agricultural Research and Development Center for the production of hydrocarbon from both guayule and sweet sorghum.

**RELATED WORDS:** biofuel

**HOW YOU'LL USE IT:** *"I hope that by the time my children are my age, all of our cars can run on something like GUAYULE or corn instead of gas."*

**G**

**Gulf Coast Project** *(guhlf KOHST proj-ekt)*, **definition**

The Keystone XL Pipeline, which carries oil from Canada into the United States, will likely be extended this year via an initiative called the Gulf Coast Project. The more than 400-mile extension plans would carry oil from depots in Oklahoma to Texas refineries along the coast. The project has been a controversial one overall, mired in objections from landowners in the pipeline's path and environmentalists who fear that the pipeline could leak, damaging waterways, wetlands, and other wilderness areas in the process. Proponents for this southern portion of the Keystone XL Pipeline say that its construction would create some 4,000 jobs. The Gulf Coast Project got the go-ahead from President Obama in 2012, and will be constructed by TransCanada. This southern portion of the Keystone Pipeline was approved quicker than other parts of the project since its construction doesn't cross an international border, which is the issue that northern extensions proposals were running into. According to the *New York Times*, the Keystone XL Pipeline "would be the longest oil pipeline outside Russia and China" and "would be able to carry more than half a million barrels a day."

**RELATED WORDS:** fracking, hydraulic fracturing, Keystone XL Pipeline

**HOW YOU'LL USE IT:** *"A couple buddies of mine are heading down to Texas to try to find work on the GULF COAST PROJECT."*

**hack** *(hak), noun*

H

One definition of the word hack—to break into, specifically via a computer or another form of sophisticated technology—is evolving as we speak. In recent years, the word has also come to mean a manipulation of sorts, of turning some preexisting thing into something new, or using one item to perform a task it's not usually intended for. There are numerous examples of this usage, with new ones emerging all the time. Take, for example, the popular notion of an Ikea hack, in which someone adapts a piece of Ikea furniture to suit their specific needs by way of some light carpentry. Home design blogs are brimming with DIYers showing off their hacking skills, adapting bookshelves into window seats or refinishing once-plain nightstands and bureaus into something unique. The word can also be applied to food, such as a recipe hack, where one ingredient is swapped out for another to accomplish the same outcome. (An example of this would be using lemon and milk when you don't have buttermilk, or molasses and sugar in place of brown sugar.) Hacks can be seen throughout the home in fact, such as with a cleaning hack, when a dryer sheet is used to thwart dust or newspapers are used in place of paper towels to help clean windows.

**RELATED WORDS:** DIY, hacker, hacktivists

**HOW YOU'LL USE IT:** *"I had no cake flour in the house so I had to do a total recipe HACK and bake the cake using regular flour combined with cornstarch instead."*

**hacker hostel** *(HAK-er hos-tl), noun*

Tied to the growth of Silicon Valley startups, the Bay Area is see-ing the development of hacker hostels—cheap, nonbinding hous-ing options for those techies living in the area who just haven't quite hit it big (yet). A July 5, 2012, article by Brian X. Chen in the *New York Times* described this trend, saying that the hous-ing attracts mostly males in their twenties, who live dorm-style in bunk beds for as little as $40 a night. The use of hacker in this term refers not to someone who is maliciously crafting ways to steal a person's identity or information, but rather to an innovator and entrepreneur in the technology and web-development start-up field. Most of these hacker hostels include a "captain"—typi-cally female, the captain serves as a sort of den mother for the crew, screening potential new residents and cooking meals. A neurosci-entist named Jade Wang is credited with helping to spread the hacker-hostel trend with her mini-chain called ChezJJ, while the rental housing website Airbnb.com has made it more mainstream. According to residents of these hacker hostels, one of the biggest benefits of the living situation is the rapid exchange of ideas and support among a group of people with common goals.

**RELATED WORDS:** Silicon Valley

**HOW YOU'LL USE IT:** *"My brother has this great idea for a new app so he moved out to California and is living in one of those HACKER HOSTELS while he works on it."*

## hacktivist *(HAK-tuh-vist)*, *noun*

A hacktivist is someone who uses Internet hacking as a vehicle for social activism or to send a political message. The word itself is a combination of hack and activism. In 2012, a group of Syrian activists allegedly gained access to the Twitter account for Al Jazeera, using it to send out tweets with an anti-revolution message. "The hack is hard to notice with a quick glance because the tweets stick closely to the news organization's usual tone," wrote Kris Holt in a post for Mashable.com. Perhaps the most famous hacktivist group right now though is Anonymous, a collective of socially motivated individuals who in recent years have taken on things like the anti-piracy act SOPA, Scientology, government agencies, and banking institutions. Anonymous has been called "a cross between an outlaw gang and a worldwide protest movement," by *New York Magazine* writer Steve Fishman. Anonymous was even named to *Time* magazine's list of "The 100 Most Influential People in the World" in 2012. Expect to hear more about hacktivists this year as more efforts by such groups develop.

**RELATED WORDS:** Anonymous, Blockupy Movement, cyberattack, Occupy Movement, slacktivist

**HOW YOU'LL USE IT:** *"I can't tell if Twitter is just down again or if it's some big HACKTIVISM ploy trying to make a social statement."*

**H**

**hen** *(hen), pronoun*

Thank Sweden, that paradigm of gender equality, for the word hen, a gender-neutral term meant to take the place of he or she. The word has roots back to the 1960s, and popped up in the 1990s during discussions by linguists about the need for a gender-neutral pronoun, and was originally meant to simply stand in for the word they. In Sweden these days, hen takes the place of han and hon—he and she respectively. The term has become a hot topic after the publication of a children's book by Jesper Lundquist used it to stand in for all gender-specific pronouns. Gender neutrality is a big concern in Sweden, where children's games are regulated and songs and books rewritten to avoid traditional gender roles. Looking forward, it's not hard to imagine the word, or at least its mission, trickling down to other languages, even English. The universal he, as in "everyone is entitled to his own opinion," is losing traction, sounding dated and sexist as time goes on. Plus, we still don't have an elegant way to refer to transgendered individuals when we're unsure of his or her sexual identity preference. (Even writing "his or her" in the preceding sentence feels awkward.) A word like hen could change that—and don't be surprised this year if it does.

**RELATED WORDS:** bridesman, cisgender, gender-neutral, groomsmaid

**HOW YOU'LL USE IT:** *"My cousins are from Sweden and rather than say his or her—as in 'the teacher reprimanded his or her student—they say 'HEN student' to stay gender neutral."*

**HI-SEAS** *(hahy-SEEZ),* *noun*

H

HI-SEAS is a NASA program that is studying how to feed a future human colony on Mars. It hopes to determine the nutritional parameters needed for such a civilization as well as work out some of the necessary resources and energy requirements while still on Earth so that astronauts can prepare and plan well before such a space mission occurs. HI-SEAS stands for Hawaii Space Exploration Analog and Simulation. This year, six astronauts from the HI-SEAS program will travel to a volcanic mountain in Hawaii, which will serve as a simulated Martian habitat for 120 days. The astronauts will spend their time utilizing only dehydrated and shelf-stable foods, such as freeze-dried cheese, beef, and chicken, as well as dehydrated fruits and vegetables. Another goal of the HI-SEAS project is to figure out a way to craft delicious meals using these products under the imposed restraints. The idea is that creating cuisine that is elevated beyond simple "space food" would serve as a link back to Earthbound society in an isolating, sensory-deprived environment like Mars. The team also hopes to address the reduced sense of smell—which affects the sense of taste—that confined astronauts encounter in places like the International Space Station. Some believe that this is a physiological coping mechanism of being cooped up amid the strong body odor that can occur. The simulation is expected to kickoff early this year.

**RELATED WORDS:** Mars

**HOW YOU'LL USE IT:** *"HI-SEAS is coming up with new ways to use space food, like using dehydrated vegetables on pizza."*

**H**

**humblebrag** *(HUHM-buhl-brag), noun*

Humblebrag is a word used to describe a statement expressing false humility while talking up one's accomplishment, wealth, status, celebrity, and so on. The humility may be sincere or feigned, perhaps done as a bit of self-mocking or a deflection of one's success. The purpose of a humblebrag (whether intentional or not) is to tell people something great about yourself. There is an entire twitter account, @humblebrag, maintained by comedian and writer Harris Wittels, whose purpose is to expose instances of this behavior. Actress Emma Watson's tweet saying, "It's been 10 years but I still feel so uncomfortable with being recognized. Just a bit shy still I suppose," is one recent example, as is former White House Press Secretary Ari Fleischer's tweet "They just announced my flight at LaGuardia is number 15 for takeoff. I miss Air Force One!!" Humblebrag culture and its prevalence as a recurring hashtag on Twitter has been gaining traction for a few years and the word was even named as one of the American Dialect Society's words of the year for 2011. Expect to hear more uses of the word as more examples naturally occur—especially with people in the public eye—and following the publication of Wittels's *Humblebrag: The Art of False Modesty*. You could even see humblebrag as a contender to be added to the *Oxford English Dictionary* this year.

**RELATED:** Facebook depression, Twitter

**HOW YOU'LL USE IT:** *"Not to HUMBLEBRAG you guys, but I can't believe someone actually bought the rights to my simple little book and might actually turn it into a movie."*

**ifttt** *(ift), acronymn*

A logic system called ifttt, which is an acronym for "If This Then That" and is pronounced like the word gift but without the g, refers to a situation where A must happen first in order for B to happen. Inversely, B can only happen if A has happened first. A simple, common application of this logical sequence is when your Twitter account has been set up to auto-tweet every time you post to Instagram. News outlets may also use ifttt so that every new blog entry posted to their website is automatically sent out via Twitter once it's published. Now, ifttt sequencing is being applied across broader platforms than just social media. Belkin WeMo devices are making use of ifttt by something called Switch, which lets you control whatever you've plugged into it by using social media, or WeMo Motion, which uses a motion sensor to control the device. According to a post by Natt Garun in Digital Trends, this could mean "getting a text when the WeMo Motion senses that the door is opening, or posting a Facebook status when someone reaches for the cookie jar." Expect to hear more about ifttt as the idea moves out of the hands of developers and into everyday technology applications.

**RELATED WORDS:** Belkin WeMo, social media

**HOW YOU'LL USE IT:** *"I'm trying to figure out a way to use IFTTT to alert me every time my roommate takes food from my side of the fridge."*

**inbox zero** *(in-boks ZEER-oh), noun*

Inbox zero may never happen for some people, while for some
it's a daily occurrence. There are those people who cannot end
their day and go to sleep at night without knowing they've accomplished inbox zero, and still more who brag about it when they
get there. Inbox zero refers to that sweet moment when all of
your new email messages have been either read, responded to, or
moved to the trash bin and what you're left with is a fresh start
and no unread emails. Achieving inbox zero may be something
to brag about to your colleagues or friends. After getting back
into the office following a vacation, it may take you hours to slog
through all of the unread messages that piled up with you were
away. "Inbox zero finally!" you may tweet or exclaim to your cubicle
mate. In this way, the difficulty implied with reaching inbox zero
can also indicate that a person is important and wanted, since only
someone who gets a lot of emails has a problem making it back
to inbox zero. Or, on the flip side of this, "One might imply that
only the most pathetic, marginalized emailer could reach such a
barren state," writes Katy Steinmetz in an online post for *Time*.

**RELATED WORDS:** connected viewer, Facebook depression,
fifth screen, FOMO, humblebrag

**HOW YOU'LL USE IT:** *"I haven't been at INBOX ZERO in
my Gmail for weeks, maybe even months—I just can't keep
up with all of the email I get."*

## Intermediate eXperimental Vehicle *(in-ter-MEE-dee-it ik-sper-uh-men-tl VEE-i-kuhl)*, *noun*

This year, a craft called the Intermediate eXperimental Vehicle (or IXV for short) will be launched from a spaceport in French Guiana for a short trip before splashing down in the Pacific Ocean. The point of the IXV's mission is to test out new technologies being used for spacecraft reentry as well as advancements in thermal protections. The vehicle is being privately built by Thales Alenia Space Italia for the European Space Agency. According to a statement on the ESA's website, "Europe's ambition for a spacecraft to return autonomously from low orbit is a cornerstone for a wide range of space applications, including space transportation, exploration and robotic servicing of space infrastructure." The IXV will have some unique features on it, specifically the absence of wings and a rounded nose. It will be launched via rocket into low orbit before descending back to Earth. Rebecca Boyle writes in *Popular Science* that "the spacecraft is more like the X-37B than the shuttle in that it will operate autonomously. It is more maneuverable and more precise than previous reusable spacecraft designs, according to ESA." Experts surmise that if all goes well, the IXV could be eventually used to transport people to the International Space Station.

**RELATED WORDS:** European Space Agency, IXV, X-37B

**HOW YOU'LL USE IT:** *"I hope they televise the launch of the IXV so we can watch it take off into space."*

**i-SODOG** *(eye-SO-dog), noun*

Built by Japanese toy company Takara Tomy, the i-SODOG is a robotic toy dog that can respond to commands—up to fifty spoken ones—and can also be controlled by a user's smartphone or a remote control. It is also equipped with artificial intelligence, which means over time it will learn the commands of its owner and even adapt its personality. The i-SODOG is small (about 4 inches × 6 inches), double-jointed, and white with LED eyes that light up in an array of colors from a bluish purple to green. In addition to doing regular doglike things such as sitting, shaking its head, barking, and lifting its leg to "pee," the i-SODOG can also dance. In the technology review online publication *The Verge*, Jeff Blagdon wrote that "there's also a Tamagotchi-like parenting component to the robot, so you can transfer his or her info to your mobile device and bring your pet with you for the day." The robot can reportedly last about an hour on a complete charge and can exchange information with other i-SODOG robots via sensors on its nose. (It takes a person to facilitate the exchange though, since touching the nose of each dog at the same time is what triggers this feature.) The i-SODOG has been compared to both Sony's AIBO robot dog and Tiger Electronic's iCybie dog, which have both been discontinued. It was unveiled at the Tokyo International Toy Show and will be available this year for about $400.

**RELATED WORDS:** AIBO, iCybie, robopocalypse

**HOW YOU'LL USE IT:** *"I'm planning to buy one of those super cool I-SODOG robots during my trip to Japan this year."*

## ISRO Mars Mission *(ISS-roh mahrs mish-uhn)*, *noun*

The ISRO Mars Mission will be launched this year by the Indian Space Research Organisation (ISRO) as a way to study the geology and climate of Mars. Another major component of the ISRO Mars Mission is to figure out if there ever was or ever could be life on Mars. The mission is especially timely as resources on Earth continue to be depleted and the conversation turns to the possibility of starting a colony outside of our planet. Some experts have surmised that the ISRO Mars Mission, once officially approved, would launch in November of 2013 when the Earth and Mars are at their closest. This close proximity of the two planets will not happen again until 2016 or 2018. As such, NASA's Mars Atmosphere and Volatile EvolutioN (MAVEN) mission will also be launching this year. The unmanned ISRO Mars Mission will be the first time India has sent a craft to the red planet and will be the furthest the country has ever sent a mission into space. It marks a big step for India as the country gets in on a space race that is often dominated by Russia, China, and the United States.

**RELATED WORDS:** HI-SEAS, Mars Atmosphere and Volatile EvolutioN mission, SpaceX

**HOW YOU'LL USE IT:** *"The ISRO MARS MISSION out of India is going to happen the same month as the United States sends a mission to Mars, creating a sort of space race between the two countries."*

J

K

L

## jamming *(JAM-ming), verb*

Jamming means to stop, prevent, or inhibit. It is most frequently associated with cell phone use. Distracted driving is on the rise in the United States and cell-phone use is to blame for 18 percent of the country's distraction-related deaths. As a way to combat this, there are now jamming devices that render a cell phone unusable while the car is in motion. The latest phone jamming device will only affect the driver's phone—everyone else in the car is able to text as much as they want. Developed by a group of Indian engineers at the Anna University of Technology, it's only a matter of time before its use becomes standard practice in cars. Previously, the devices used for jamming were controversial since misuse also inhibits radios and telephones used by police officers and other public-safety officials. In addition to these jamming devices, there is also something called a SpeechJammer gun, invented by Kazutaka Kurihara of the National Institute of Advanced Industrial Science and Technology and Koji Tsukada of Ochanomizu University. The device is used to record a person's speech and play it back to them at a .2 second delay. Hearing your voice echoed in such a way is extremely disorienting, and negatively affects the brain's cognitive process, rendering its "victim" speechless.

**RELATED WORDS:** distracted driving

**HOW YOU'LL USE IT:** *"I'm getting my mom a JAMMING device for Mother's Day so I know she'll stop talking on her phone when she's driving."*

**kidsick** *(KID-sik), noun*

You know the word homesick—when kids go off to camp or somewhere else away from home and miss their parents—but what about when it's the other way around? The idea of kidsick, described by Bonnie Rochman in *Time* magazine, is when parents miss their kids while they're away from home. As such, a new trend is developing at sleepaway camps where staff members update parents with photos of their kids at camp via Facebook, the camp website, even e-mails. Writes Rochman, "camp used to be a place kids went to learn self-reliance and discover themselves away from the watchful eyes of mom and dad, but now technology is allowing parents to keep tabs on their kids even from afar." In the article, CEO of the American Camp Association Peg Smith is quoted as saying, "It's not the kids who are homesick. It's the parents who are kidsick." Our digital, plugged-in culture and the prevalence of helicopter parenting are feeding this kidsick behavior. Expect to hear more about it as camps and even schools grapple with how to appease and abate kidsick parents in the digital age.

> **RELATED WORDS:** attachment parenting, free-range kids, helicopter parenting, smother mother
>
> **HOW YOU'LL USE IT:** *"When the twins went off to camp I was just so KIDSICK I started logging into Facebook every day to see if anyone had posted photos of them while they were there."*

**Kunstkammer Vienna Museum** *(KOONST-kahm-ur vee-EN-uh myoo-zee-uhm),* **noun**

After being closed for nearly a decade, the Kunstkammer Vienna Museum will reopen in March. The Kunstkammer, which is generally translated as a cabinet of curiosities, art room, or cabinet of wonder, includes various exotic objects, sculptures, pieces of art, and quirky items. There are things like ostrich eggs, a saltcellar, gameboards, clocks, strange scientific instruments, and then the horn of a "unicorn." Part of the Kunsthistorisches Museum, the Kunstkammer has more than 2,200 objects displayed in a newly renovated space that measures 29,000 square feet. According to an Austrian travel website, "the collection is regarded as a reflection of the entire universe, its task to transmit knowledge and amaze all who see this fabulous realm of fantasy and curiosity." The objects housed within date back to the time of the Habsburg emperors, when it became popular to collect such rare and unusual objets d'art made specifically for the aristocracy. Collections such as this can cover wide-ranging genres like geology, art, religion, and history, and often include fantastical items and curiosities, such as the Kunstkammer's unicorn horn—most likely the tusk of a Narwhal.

**RELATED WORDS:** art, curiosities, museum

**HOW YOU'LL USE IT:** *"We are extening our Europe trip to Austria now that we learned that the KUNSTKAMMER VIENNA MUSEUM will be reopening just before we're there."*

**laser ablation** *(ley-zer a-BLEY-shuhn)*, *noun*

As scientists and astronomers continue to prepare for an asteroid on a collision course with Earth, talk of things like laser ablation continues to pop up. We'll get a close visit from the 2012 DA14 Asteroid this year, thus prompting another round of discussion about just how to deflect such a thing someday. Laser ablation is a method of asteroid defense that would involve using solar-powered lasers directed on the asteroid, which would cause the surface to sublimate and transform from a solid to a gas (and skip the liquid phase in between.) According to a story in *Technology Review*, the theory behind laser ablation "is that the material vapourised from the asteroid's surface, pushes it like rocket exhaust, generating thrust." Even if the asteroids are out of range, the theory is that laser ablation could still knock an asteroid off course. According to a post by Ariel Schwartz on the *Fast Company* website, the advantage of this solar-powered technique is that "they require no outside fuel source, there's more room for error if one of the lasers doesn't do the job, and they're cheaper to cool."

**RELATED WORDS:** 2012 DA14 Asteroid, B612 Sentinel, gravity tractor

**HOW YOU'LL USE IT:** *"It sounds like something you see in the movies but LASER ABLATION could potentially be used to save us from an asteroid that would destroy us all."*

**L**

## LightSail Energy *(LAHYT-seyl EN-er-jee)*, *noun*

Founded by a young scientist named Danielle Fong, the Berkeley-based LightSail Energy is devising ways to store power we're not using and pump it back into the grid. The method used to store the energy is by way of air compression—when a windmill or solar panel generates electricity, it powers a compressor and puts air in the tank. When you need to use the energy later, the air is released and drives the generator. Fong and her team helped to create a method for doing this that involves spraying water into the air so the temperature stays consistent and the pressure stays high. According to a post by Whitney Pastorek on the *Fast Company* website, "LightSail's process recovers 70% of the energy it puts out, pretty much doubling the efficiency of the standard compression method." The technology is expected to go into practice sometime this year when LightSail Energy starts delivering the units, which are as big as shipping containers.

**RELATED WORDS:** fuel cell, renewable energy, smart grid, tidal energy

**HOW YOU'LL USE IT:** *"After we get married we're planning to build a solar-powered house and are hoping we'll be able to use a unit by LIGHTSAIL ENERGY so we have a good way to store what we don't use."*

## LimbIC *(LIM-bik), noun*

Could the $8,500 LimbIC chair be the next trend to overtake office culture? In the last few years, we've seen people sitting on yoga balls at their desks, and now standing up while working at their computers. It's clear that people are looking for a way to break the sedentary tradition and work in an ergonomic position that is better for their bodies. Enter LimbIC, an innovative chair designed by Dr. Patrik Künzler and his Zurich-based company Inno-Motion following Künzler's time at MIT. Rather than sitting on a flat seat, someone using the LimbIC chair perches atop a pair of carbon-fiber shells that wrap snugly around each thigh. Sitting in the chair allows you to move around, twist, even dance if you're so inclined. The chair's design is meant to feel like an extension of the body, keeping the muscles engaged so that you're never quite sitting still. Künzler claims that his chair design helps to encourage a person's creativity while they're working and to produce feelings of happiness. The design of the chair also helps to lubricate the spine and improve circulation, all while keeping your feet suspended and off the floor.

**RELATED WORDS:** cubicle, ergonomic, limbic system, office culture

**HOW YOU'LL USE IT:** *"Our company is getting us all those LIMBIC chairs because they heard that they'll make us happier and more creative, plus keep us moving so we stay active."*

**L**

### Lipstick Effect *(LIP-stik ih-fekt), noun*

During times of economic duress women are more likely to spend money on cosmetics or beauty products as a sort of quick pick-me-up. The Lipstick Effect is the catchy name given to this phenomenon, which was determined through a series of studies conducted by researchers at Texas Christian University, the University of Minnesota, the University of Texas at San Antonio, and Arizona State University. Researchers concluded that when reminded of hard economic times, women were more interested in purchasing beauty products than other items. (Their male counterparts, however, did not show an increase in the desire to buy grooming products after they'd been reminded of the recession.) The researches also found that advertisers capitalize on hard times by marketing their cosmetics to women specifically as a way to attract men. "The paper's authors theorize that since the supply of financially stable men falls during recessions, the price that women choose to pay for their attention spikes," writes Bonnie Karoussi in an article about the study on the *Huffington Post*. Leonard Lauder, the chairman of Estée Lauder, is credited with first coining the phrase after his company noticed an uptick in lipstick sales following the events of September 11, 2011.

**RELATED WORDS:** the Leading Lipstick Indicator

**HOW YOU'LL USE IT:** *"Maybe it's the LIPSTICK EFFECT, but I seriously can't walk into a CVS without buying some new nail polish or eye shadow these days."*

**LiquiGlide** *(LIK-wi-glahyd), noun*

It won't be long before LiquiGlide technology is coating the insides of every bottle and can, rendering the long-performed ritual of shaking a bottle of ketchup a thing of the past. Developed by a team at MIT, the LiquiGlide project has created a coating that can be sprayed inside of a bottle before it's filled, rendering its surface super slippery and helping its contents to simply pour right out. Substances such as mayonnaise and ketchup have been used to display LiquiGlide's properties, which allows them to pour out just as easily as a nonviscous liquid like water would. It would also help reduce food waste, since it doesn't allow for anything to be left stuck to the inside of the bottle. In an NPR interview with doctoral MIT student Adam Paxon who helped develop LiquiGlide, Paxon is quoted as saying, "we've gotten a lot of interest from a bunch of food companies. Any time you have a really thick sauce that's hard to get out of the bottle, this helps to let it slide." The kicker about LiquiGlide technology is that it's an edible substance and that all of its individual components are already FDA-approved.

**RELATED WORDS:** WikiCells

**HOW YOU'LL USE IT:** *"I wish ketchup bottles were coated with LIQUIGLIDE back when I was waitressing so we didn't have to help the customers shake out the ketchup all the time."*

**L**

## locaflor *(LOH-kuh-flohr), noun*

As you worry about your carbon footprint and try to do a better job of eating locally, the thought starts to arise: what else should you be buying locally? Locaflor, which combines the words local and flora, refers to a growing trend to buy flowers that have been grown locally. The benefits to buying local flowers are numerous. For one, it's better for the environment and cuts out the carbon emissions that happen when flowers are forced to travel a long distance from grower to buyer. Making sure that flowers last during this time also requires the use of chemicals and refrigeration. When flowers are grown and purchased locally, that just isn't the case. *Freakonomics* coauthor Stephen Dubner helped shed some light onto the economic and environmental side effects of buying flowers that have been grown far away and shipped to your door. During an appearance on NPR's *Marketplace* in 2012 around Mother's Day, Dubner told host Kai Ryssdal that we spend about $12 billion every year in the United States on cut flowers, 80 percent of which are imported. Dubner tried to make the argument on air that the solution to this is to instead send fake flowers. What's more likely though is a movement geared toward the locally grown. Expect to hear more about this this year as the locaflor trend really starts to take off.

**RELATED WORDS:** carbon footprint, food miles, foraged, locavore, merroir

**HOW YOU'LL USE IT:** *"My fiancé and I are really worried about the environmental side effects associated with having flowers from far away, so we're doing a total LOCAFLOR wedding."*

**Lufa Farms** *(loo-fuh FAHRMZ), noun*

Montreal-based Lufa Farms, an urban farm group founded by Mohamed Hage and Kurt Lynn, are expanding out of Canada and into the United States this year. The idea behind Lufa Farms, and other urban rooftop farms like it, is pretty simple: put a greenhouse on the roof of a building and extend the growing season well past the warm summer months. Then sell the produce to consumers who live close by. It's a whole new take on the locavore movement, injecting the philosophy into an environment where agriculture isn't typically found nearby. Lufa Farms isn't the only one to get in on the urban rooftop-gardening trend, but they're likely to be the one we're all talking about this year as they explore expansion options in Boston, New York, and Chicago. The goal is to have fifteen to twenty U.S. locations in operation by 2020, plus six more in Canada. In addition to the obvious benefits of fresh produce, urban rooftop farms like Lufa Farms also help lower the energy costs for building owners.

**RELATED WORDS:** agribusiness, BrightFarms, Gotham Greens, hydroponics, urban rooftop farming, vertical farming

**HOW YOU'LL USE IT:** *"I hope LUFA FARMS comes to my city and builds a greenhouse on the roof of my office building so that I have an easier way to get fresh produce."*

M
N

**M**

**MaKey MaKey** *(MEY-kee MEY-kee)*, *noun*

MaKey MaKey is a highly adaptable device that can turn just about anything into a keyboard or controller as long as it can conduct electricity. The name combines the words make and key, as this is exactly what you're doing with the product. MaKey MaKey was funded with help via Kickstarter, and after the $25,000 goal was surpassed, the project went on to earn a total of $568,106 in just thirty days, a clear indication that the product is sure to be a hit once it goes on the market. MIT PhD students Jay Silver and Eric Rosenbaum are the masterminds behind MaKey MaKey, which can do things like turn bananas into a piano, Play-Doh into a keyboard, or an apple into a computer mouse, all simply by hooking these items up to the device using alligator clips. "MaKey MaKey isn't just a tool and a toy—it's a new set of working assumptions about what it means to own and interact with technology," writes John Pavlus in a post on the website for *Fast Company*. The idea is to create the ultimate user-interface experience, making a controller and application that makes sense to you.

**RELATED WORDS:** crowdfunding

**HOW YOU'LL USE IT:** *"Once I get a MAKEY MAKEY I'm going to see how many different things I can use from my fridge to make a computer mouse."*

**mancession/hecovery** *(man-SESH-uhn / hi-KUHV-uh-ree)*,
*noun*

M

Though the term mancession was first coined in 2008 to refer to
the fact that job loss and the recession was hitting men harder
than women, it's popping up again amid discussion of economic
recovery. Or, as some flippant journalists and economists are calling it: the hecovery. Though men may have been hit hardest by
unemployment during the recession, they are now finding jobs
faster in the recovery. Don Lee writes in a story for the *LA Times*,
"the gender gap has raised concerns about possible discrimination
in hiring. If the trend persists, it could set back gains made by
women in the workplace, experts said." The word mancession is
a combination of the words man and recession. Many have surmised that a mancession comes about because manufacturing jobs
and construction—both male-dominated fields—take the biggest
hit when the economy tanks. In the last few years, though, it
appears that things have been turning around. One of the reasons
given for the hecovery (that's he + recovery), or the mancession
reversal, is that during tough times, these same men took jobs that
are normally favored by women, such as in retail.

**RELATED WORDS:** double-dip recession, economy,
Generation Debt

**HOW YOU'LL USE IT:** *"All of the guys I know who were out
of work a couple of years ago during the MANCESSION
are now back working and rebuilding their finances—thank
goodness for the hecovery!"*

**M**

**manimony** *(MAN-uh-moh-nee)*, *noun*

Manimony is alimony paid to a man. Traditionally, it's been common for the man to be the one to pay alimony to the woman following a divorce, since in many cases the man is making more money, or is even the sole breadwinner while the woman has been working a less-demanding job and raising the children. However, now that those social norms have been flipped (and especially due to the proliferation of high-profile divorces among celebrities) manimony is becoming more common. Though the concept of women shelling out big money during divorce settlements isn't something new, the word manimony is. A combination of the words "man" and "alimony," the use of manimony is on the rise as the trend continues to grow. Anderson Cooper did a special segment called "The Man-imony Controversy" on his talk show in 2012 about the need for alimony reform in the United States. Among those splits that could see a major payout this year include the divorces of Courtney Cox and David Arquette, Debra Messing and Daniel Zelman, and Kris Humphries and Kim Kardashian.

**RELATED WORDS:** breakover, grey divorce

**HOW YOU'LL USE IT:** *"Well, since Meredith has been the one pulling in the bigger salary the past few years, it looks like she's going to have to pay Pete a decent amount of MANIMONY once their divorce is final."*

**marine drone** *(muh-REEN drohn)*, **noun**

M

A marine drone is a pollution-battling device designed by Elie Ahovi in response to a challenge by French environmental firm Veolia. Veolia asked design students to create something to help combat problems with trash in our oceans, specifically the Great Pacific Garbage Patch, a collection of trash that has amassed in a remote part of the Pacific Ocean that is roughly twice the size of Texas. Ahovi's marine drone is autonomous, meaning it could travel through the water unmanned, while sucking in trash through an opening and collecting it in a net, which it drags behind. It would be able to collect everything from microplastic shards to larger items like plastic bottles and would be able to troll the sea for two weeks before returning to its home base to drop off the trash. "It discourages fish and other creatures from entering its jaws via an annoying sonic transmitter, and it communicates with other drones and with its base station using sonar," writes Rebecca Boyle on the website for *Popular Science*. The development of such a device would come at a crucial time as it is expected that large amounts of debris from the 2011 Japan tsunami will join the patch this year.

**RELATED:** Great Pacific Garbage Patch, microplastic, Pacific Trash Vortex, Project Kaisei

**HOW YOU'LL USE IT:** *"I would love to see the widescale use of MARINE DRONES get some funding so that we can clean up our oceans once and for all."*

**M**

## Mars One *(mahrz WUHN)*, *noun*

Based in the Netherlands, a privately owned space exploration venture called Mars One hopes to set up a Martian space colony and send four astronauts to live there in 2022 (arriving in 2023), plus a new pair to join them every two years. This ambitious plan for a colony on Mars is the brainchild of Dutch entrepreneur and researcher Bas Lansdorp, who plans to fund Mars One by what he calls a "media spectacle." Some have surmised that this could mean corporate sponsorships or broadcasting rights. Perhaps we're not that far away from *The Real Word: Mars*? The timeline for Mars One is very specific in terms of dates, with the first goal set for 2016 when Mars One plans to send a communications satellite to the planet. Following that will be a rover in 2018 and then the building infrastructure in 2020, which could include things like solar panels. These next few years are certainly shaping up to be prime time for private space companies such as this. The exploration of Mars and the consideration of the planet as the location for a potential human colony are also being looked at by the ISRO Mars Mission this year.

**RELATED WORDS:** HI-SEAS, ISRO Mars Mission, Mars Atmosphere and Volatile EvolutioN mission

**HOW YOU'LL USE IT:** *"If MARS ONE is successful, we could see people living on the red planet within our lifetime."*

**MAVEN** *(MEY-vuhn), acronym*

M

MAVEN stands for Mars Atmosphere and Volatile EvolutioN, a NASA program that sent a probe to Mars in 2012. The goals of the MAVEN mission involve the study of the planet's atmosphere and ionosphere, as well as a look at how solar winds interact with the planet. The project, which is being spearheaded by Bruce Jakosky of the Laboratory for Atmospheric and Space Physics at the University of Colorado at Boulder, also intends to take a look at what past conditions on Mars could have been like. It is believed that there was once water on Mars as well as a thicker atmosphere, due to the discovery of some dried-up riverbeds on the red planet and some minerals that are generally found when there is water present. Scientists are trying to figure out what could have happened to change the environment on Mars and if it could ever be hospitable to humans. MAVEN is just one of many programs and projects with an eye on Mars, several of which are slated for this year—there is also India's ISRO Mars Mission, Mars One, and the Mars simulation project HI-SEAS.

**RELATED WORDS:** Curiosity rover, HI-SEAS, ISRO Mars Mission, Mars One, VASIMR

**HOW YOU'LL USE IT:** *"Wouldn't it be so weird if the MAVEN mission went to Mars and discovered that there was already life there millions of years ago and that the fate of our planet is to become like Mars?"*

**M**

## MEG 4.0 *(MEG-fohr-point-oh), noun*

Augmented-reality glasses will be a big trend this year thanks to Google's Project Glass and the MEG 4.0 from Olympus. The 30-gram, wearable computer works via a Bluetooth connection to a smartphone, which enables any pair of glasses to display content on a tiny screen within the user's field of vision but without actually obstructing his or her view. The battery life is reportedly eight hours for MEG 4.0 when users are casually tuning in for short bursts of use—the device would be able to last for two hours if it is being used to project nonstop. When news broke of MEG 4.0 in 2012, it was still just in its prototype phase. Notably absent from MEG 4.0 were camera capabilities for both still shots and video—both are features that the Google glasses have. Though Olympus hasn't copped to a specific release date or price point yet, it's easy to see that the proliferation of these wearable computing devices in the coming years will have everyone talking.

**RELATED WORDS:** augmented reality, Facebook depression, fifth screen, FOMO, Google Glasses, Project Glass

**HOW YOU'LL USE IT:** *"I'd be more inclined to buy the MEG 4.0 when it comes out since I already wear glasses and could just hook it onto my prescription pair."*

**merroir** *(mer–WAHR), noun*

The locavore movement has helped get everyone talking about where they are getting their food, with an emphasis on local and fresh. It helped pave the way for other food trends, like the farm-to-table movement, as well as spark discussions about things like merroir. Think of merroir as the seafood equivalent of terroir. Terroir is a French word, which roughly translates to soil, region, or local—something which is specific to a particular location. (For example, a "recette du terroir" is a local recipe.) In winemaking, the terroir refers to the condition and characteristics of the soil in which the grapes are grown. Things like soil type, nearby plants, and climate can all contribute to the terroir. Since the terroir can differ from place to place, it contributes to the unique characteristics in various wines. Likewise, the merroir is the conditions in which seafood is harvested or produced, specifically in reference to oysters. It is the "sense of place" for shellfish. Merroir is also the name of a Virginia restaurant, opened in 2011, which seeks to highlight a variety of oysters from Rappahannock River Oysters. Owners and cousins Travis and Ryan Coston, along with chef Peter Woods, are widely credited with jump-starting the idea of merroir. Erin Byers Murray's book *Shucked: Life on a New England Oyster Farm* features this trending term, as do a range of food blogs.

**RELATED WORDS:** farm to fork, farm to table, locavore, terroir

**HOW YOU'LL USE IT:** *"You can really taste the difference in the MERROIR when you eat an oyster from Massachusetts versus an oyster from Virginia."*

**M**

## miCoach *(MAHY-kohch), noun*

The widespread use of the Adidas miCoach by Major League Soccer was announced in the summer of 2012 at the New Museum of Contemporary Art in Manhattan and will go into use by every MLS player starting in the 2013 season. The miCoach is a super high-tech sports monitoring device that fits into a small pocket between a player's shoulders. The gadget uses sensors to monitor a player's performance and then beams that information back to a human coach who can check out the data in real time on an iPad or reference it later to help analyze a game. The technology inside each device includes GPS (to track how the player is covering the field), a magnetometer, and a gyroscope. In addition to the device, each player's base-layer shirt will also be woven with tiny sensors that can monitor vitals like heart rate. For the time being, the data will only be visible to the players' coaches, but down the line it could be something that might be used as a talking point during the broadcast of a game and its ensuing analysis. A wide range of miCoach products are already available for sale to nonprofessional sports enthusiasts looking to monitor their performance.

**RELATED WORDS:** FitBit

**HOW YOU'LL USE IT:** *"There's really no way for MLS players to fake that they're tired if their coach is monitoring their every vital with the MICOACH."*

**microplastic** *(MAHY-kroh-plast-tik), noun*

M

Environmental woes, especially those related to pollution and global warming, will be big concepts for this year as we continue to struggle with how to take care of our planet. One term you will hear in relation to a surge in marine pollution (due in part to refuse from the Japan tsunami washing out to sea) is microplastic. The majority of trash lost at sea is broken down into rice-sized bits or smaller, called microplastic. These small pieces of plastic are part of a big problem that can affect our food chain. Microplastics pose a big risk to fish, which eat the chemical-laden bits and can't digest them. This is also risky for you, since you eat these fish that are full of toxins. Microplastics can also refer to even smaller, fibrous bits of plastic that are shed off of synthetic clothes when they're washed and are then leached back into the food chain. Microplastics are often associated with the Great Pacific Garbage Patch, a swirling mass of trash that's roughly twice the size of Texas.

**RELATED WORDS:** Great Pacific Garbage Patch, marine drone, pollution

**HOW YOU'LL USE IT:** *"I've decided that from now on I'm only going to wear cotton, wool, or other natural-fiber clothing so that when I wash my clothes I'm not leaching MICROPLASTICS into the food chain."*

**M**

**Microsoft Surface** *(mahy-kroh-sawft SUR-fis), noun*

Microsoft will be making a big leap from software to hardware soon with the development of a tablet called Surface, which is widely considered to be a direct competitor to Apple's iPad and even its MacBook Air. The Surface tablet will showcase a variation of Microsoft's new Windows 8, which has been optimized for touchscreen devices. The Surface will have a built-in kickstand, so that it can be propped up like a laptop, as well as a detachable keyboard (in colors like pink, blue, and charcoal) that doubles as a screen cover. The screen itself will measure 10.6 inches. Microsoft debuted the product to the press in June of 2012, announcing the event with just a few days notice, much in the same way that Apple debuts its latest products. According to a story by Nick Wingfield in the *New York Times*, "with the detachable keyboard for Surface, known as Touch Cover, Microsoft seemed to be positioning its tablet as a more business-friendly alternative to the iPad, one that is better suited to productivity tasks that require faster typing."

**RELATED WORDS:** Apple, fanboy/fangirl, fifth screen, iPad, tablet

**HOW YOU'll USE IT:** *"I can't wait for the MICROSOFT SURFACE to come out—I'm going to be first in line to buy one on the day it goes on sale."*

## Milkomeda galaxy *(milk-OM-i-duh GAHL-aks-see)*, **noun**

M

Scientists are starting to talk about the development of a new galaxy—our own—when the Milky Way and Andromeda inevitably collide in some 4 to 5 billion years. A combination of the names of the two galaxies has led scientists to call the result of the inevitable collision Milkomeda. Though it's quite some time away, the possibility of a collision between the two galaxies has been discussed for years. In 2012, researchers using the Hubble Telescope were able to confirm that it will indeed happen someday, making it a hot topic of discussion this year. When it happens, the sun's position will likely be relocated, the shape of the galaxy will change, and our view of the nighttime sky will be altered, though scientists say Earth is not in any danger of being destroyed. According to a story in the *Guardian* about the collision, it is unlikely that any planets or stars will collide since "the space between them is equivalent to a football field between grains of sand."

**RELATED WORDS:** Andromeda galaxy, Hubble Telescope, Milky Way galaxy

**HOW YOU'LL USE IT:** *"The craziest thing for me in terms of the MILKOMEDA GALAXY is the fact that all of the stars in the sky will look completely different from how they look now."*

**M**

**mullet dress** *(MUHL-it dres), noun*

One fashion trend sure to be popping up through the year is the mullet dress (or skirt). The hemline of a mullet dress is shorter in the front than it is in the back, getting its name from the oft-derided mullet haircut that features shorter hair on the front and top of the head and longer hair in the back. ("Business in the front and party in the back" is the common phrase associated with the mullet hairstyle.) The mullet-dress trend, featuring what is also sometimes referred to as a "high-low" dress or skirt, started in 2012 and is continuing into 2013. Thank celebrities and fashion designers for perpetuating the trend on the red carpet, such as the long-sleeved black mullet dress by Emilo Pucci that Nicole Richie wore to an awards show in 2012. Stars like Robyn, Blake Lively, Selena Gomez, and Chelsea Handler have all been spotted sporting this look, too. It should be said, though, that a mullet dress is a difficult look to pull off for many, often landing its wearers in the worst-dressed lists. Expect to see more mullet dresses this year as the trend that was everywhere on the runways of 2012 trickles down to the sale racks.

**RELATED WORDS:** designer, fashion, haute couture, runway, sale rack

**HOW YOU'LL USE IT:** *"I saw the cutest polka dot MULLET DRESS on a girl walking down the street the other day and I'm kicking myself for not asking her where she got it."*

**myelodysplastic syndrome** *(mahy-loh-dis-PLAS-tik dis-AWR-der), noun*

**M**

A rare blood disorder called myelodysplastic syndrome, or MDS for short, became a headline in 2012 after *Good Morning America* host Robin Roberts publicly announced her diagnosis. MDS causes stem cells in a person's bone marrow to become defective blood cells (rather than healthy ones), which eventually outnumber the healthy ones in the blood stream. Symptoms during the early stages include fatigue, bruising, and a shortness of breath. Exposure to chemotherapy and radiation can lead to MDS. In the case of Roberts, the ABC TV personality was diagnosed with, treated for, and cleared of breast cancer five years prior to her MDS diagnosis. Treatment for MDS often includes blood transfusions and bone marrow transplants. Roberts announced on *Good Morning America* that her sister, Sally-Ann Roberts, would be her donor. In a first-person post by Roberts on the *Good Morning America* website, the newscaster wrote, "If you Google MDS, you may find some scary stuff, including statistics that my doctors insist don't apply to me. They say I'm younger and fitter than most people who confront this disease and will be cured." Discussion about MDS will surely continue through 2013 as Roberts receives treatment and speaks to media outlets about the disease.

**RELATED:** bone marrow transplant, cancer, chemotherapy, radiation

**HOW YOU'LL USE IT:** *"It seems like a cruel hand to be dealt that after Robin Roberts was cleared of breast cancer she was diagnosed with MYELODYSPLASTIC SYNDROME."*

**N**

**Nautilus Minerals** *(NAWT-l-uhs min-er-uhlz)*, **noun**

Toronto-based company Nautilus Minerals is an underwater mining company that is planning to undertake the first deep-sea mining project in the world (called Solwara I) starting this year. The company will use machinery and technologies common to the oil and gas industries, which have been adapted to drilling in the ocean floor. The mining will happen off the coast of Papua New Guinea in the Bismarck Sea, some 1,600 meters beneath the water and withstanding pressure of more than 1.5 tons per square inch. The purpose of the Nautilus Minerals drilling project is to extract copper and gold from a seafloor massive sulfide deposit (SMS) for the first time. Other known SMS deposits are located in the Colville Ridge and the Kermadec Volcanic Arc in New Zealand. In the years to come, Nautilus Minerals will continue its seafloor exploration in search of copper, gold, zinc, silver, and more SMS deposits.

**RELATED WORDS:** deep-sea mining, fracking, marine drone, seafloor massive sulfide, tidal energy

**HOW YOU'LL USE IT:** *"Some people say it's not space that's the final frontier, but the ocean floor—something the underwater mining company NAUTILUS MINERALS knows all too well."*

**Next-Generation Identification** *(nekst-jen-uh-REY-shuhn ahy-den-ti-fih-KAY-shun), noun*

N

Biometric technology isn't just restricted to the movies anymore. In fact, the FBI is getting in on this sophisticated technology by way of the Next-Generation Identification system. Biometrics refers to the unique traits that a person possesses, which can be used to identify them. Things like voiceprint, fingerprints, DNA, and a person's irises are all examples of biometric identifiers. The $1 billion program is extending the FBI's capabilities for identifying a person, building on their fingerprint database to include things like facial images, palm prints, and even iris scans. The Next-Generation Identification system, or NGI for short, is expected to systematically replace the FBI's current Integrated Automated Fingerprint Identification System. According to the FBI's website, "the framework will be expandable, scalable, and flexible to accommodate new technologies and biometric standards, and will be interoperable with existing systems." The FBI plans to start testing their database by next year, so expect to hear about it in the time leading up to its official launch.

**RELATED WORDS:** biometric ATM, biometrics, FBI

**HOW YOU'LL USE IT:** *"My cousin is a police officer and she told me that the NEXT-GENERATION IDENTIFICATION SYSTEM will help them at work because when they're scanning someone's iris to get it into the database, they don't have to make contact with them the way they do with fingerprinting."*

**N**

**night market** *(NAHYT mahr-kit), noun*

A new, emerging trend on the food, dining, and nightlife scene is the idea of a night market. Like the name suggests, these outdoor assemblies of food vendors and retailers happen at night and often involve shutting down streets to vehicles for the occasion. Popular in Asia, where events such as these are a mainstay, the night-market trend is catching on in places like San Francisco, San Jose, Pasadena, and other parts of California. Taking its cues from Asia, the California night markets have been a way to showcase pan-Asian cuisine. These nocturnal events frequently feature musical entertainment as well as vendors and are a place to shop, dine, and socialize. The street-food trend in the United States has been picking up in the last few years, with more cities granting food-truck permits and organizing festivals such as this. As is the case with many trends, what starts in California often makes its way across the country not long after. As such, don't be surprised to see night markets pop up in other cities across the country this year.

**RELATED WORDS:** bazaar, farmer's market, shopping

**HOW YOU'LL USE IT:** *"We got to check out this great NIGHT MARKET during our trip to San Francisco last week, where we sampled all different kinds of amazing Asian food."*

## Nomiku *(NO-mee-koo), noun*

N

Thermal immersion cooking, which involves cooking something in a vacuum-sealed bag (or some other airtight container) that is then submerged in water, has long been a practice of top chefs and culinary experts. The technique, which is also called sous-vide, can often be seen on popular cooking competition shows like *Top Chef* and *Iron Chef America*. Nomiku is a $359 device that takes the practice of sous-vide into the kitchen of home cooks. The development of the device, which is about the size of a hand blender, received funding by way of the popular crowdfunding website Kickstarter. The purpose of sous-vide cooking is to continuously heat something at a precise temperature, something that is achieved by using an immersion circulation. For example, a medium-rare steak cooks through at 57 degrees. If you were cooking steak using the Nomiku, you would set the device to 57 degrees and let it work its magic. The handy thing about the Nomiku is that it can do this in any pot that you already own, so there's no need to buy more equipment. Pre-orders for the Nomiku will have gone out by the end of 2012, so expect to hear more about this innovative cooking product this year.

**RELATED WORDS:** cooking, crowdfunding, foodie, gastronomy, immersion circulator, sous-vide

**HOW YOU'LL USE IT:** *"You will not believe how much easier it is to make custard with the NOMIKU—it's so simple now and requires so little work on my part."*

# O
# P
# Q
# R

## oblication *(ob–li–KEY–shuhn)*, *noun*

The word oblication is a portmanteau coined to give a term to a growing issue in the United States. An oblication is an obligation vacation in which someone must eat into their allotted vacation time at work in order to have time off to do something they need to do, like go to a wedding. A 2012 story on CNN.com detailed this growing trend, which affects more than 40 percent of U.S. adults every year according to a survey conducted by the popular travel-booking service Hotwire. In her story for CNN, Marnie Hunter interviews sociologist Jeffrey Alexander, who says, "On the one hand, doing these rituals is a way of showing that you're not a materialistic person and that you value your close friendships." He goes on to say, "At the same time, we do live in the economic world." For workers with very little vacation time already, the oblication can be frustrating. Sure, weddings are fun and certainly an important occasion that brings a couple's family and friends together for their special day. But for people in their twenties and thirties—when vacation time is scarce and still being accrued, and the rate of friends' weddings is high—the oblication can also be bittersweet, since it cuts into time that they may have wanted to use for something else. Like a vacation.

**RELATED WORDS:** destination wedding, social obligation, social pressure, vacation time, wedding

**HOW YOU'LL USE IT:** *"I always get so annoyed when I have to take time off to go to a wedding and my boss tells me to enjoy my vacation—um, hello, it's an OBLICATION!"*

**ocean acidification** *(OH-shuhn uh-sid-uh-fuh-KEY-shun)*, *noun*

**O**

The use of fossil fuels, and thus the release of greenhouse gases like $CO_2$, is causing the pH of our planet's oceans to change. Specifically, the oceans are becoming more acidic, something which is referred to as ocean acidification. The change to the chemistry of our seas has been referred to as "global warming's evil twin." The hazards posed by a greater increase in ocean acidification are tremendous. It would mean that animals like crabs, lobsters, shrimp, and various other sea creatures that frequent our dinner tables wouldn't be able to build their shells. The areas affected the most by ocean acidification include the coasts of Washington, Oregon, and California in the Pacific Ocean and Portugal and Africa in the Atlantic. Expect to hear more about ocean acidification as scientists continue to study its effects and devise ways to try to thwart it. We may also hear about this process during talks regarding environmental policy.

**RELATED WORDS:** carbon capture, geoengineering, global warming

**HOW YOU'LL USE IT:** *"My cousin's shrimp business could be in big trouble in a couple decades if OCEAN ACIDIFICATION continues the way researchers are predicting."*

**O**

**offline dating** *(awf-lahyn DEY-ting), verb*

At first glance, offline dating seems like it would simply refer to the opposite of online dating. Where online dating requires you to create a profile and get matched up with potential suitors via the Internet, offline dating sounds like it would mean dating in the real world. And in a sense it does, except offline dating has come to refer to a situation that has been fabricated by an online dating site, but is indeed happening offline. Take for instance the dating website How About We, in which one user proposes a date (as in "How about we go see that new blockbuster and eat lots of popcorn") and another can respond. Members still upload photos of themselves and essentially create an online dating profile, but the emphasis of the site is to get people out into the real world and on an offline date as soon as possible. Popular online dating website Match.com is also getting into the offline dating game with the creation of The Stir, a series of events hosted by the website that invites members out to meet in a real-life, offline setting completely outside of their site. It's just the latest example of offline dating, which is something we're likely to see more of this year as the trend heats up.

**RELATED WORDS:** mixer, online dating, relationship

**HOW YOU'LL USE IT:** *"I'm more into joining a site that has OFFLINE DATING events since I would really feel weird telling people that I met my girlfriend online."*

**OpenROV** *(OH-pun ahr-oh-vee), noun*

The folks behind the San Francisco-based OpenROV, who call themselves "DIY Ocean Explorers," took to crowdfunding site Kickstarter as a way to pay for their project. The idea behind OpenROV is pretty simple: anyone can build it and anyone can become an ocean explorer. The shoebox-sized robot is designed to be low cost, made with off-the-shelf parts, and assembled at home. It can travel to a depth of 20 to 100 meters, where it can explore via remote control and send images to its operator in real time on the web. NASA engineer Eric Stackpole is the brains behind OpenROV, which includes headlamps and a remote-controlled camera. It is powered by eight C batteries and runs best in fresh water, though developers are working on ways to better adapt OpenROV to salt water. They hope that initial testing and feedback from their Kickstarter investors will help them to make the necessary changes. OpenROV, which weighs just 5.5 lbs, is truly an open source project—thus the name—and as such its founder will be looking for that all-too-important feedback.

**RELATED WORDS:** Aquarius Reef Base, crowdfunding, DIY, Google Lunar X-prize, ocean acidification

**HOW YOU'LL USE IT:** *"I can't wait to get my OPENROV kit in the mail as a project that my son and I can work on together."*

**O**

## Operation Twist *(op-per-ray-shun TWIST)*, *noun*

Operation Twist is a Federal Reserve program that went into effect in 2011, which attempts to lower interest rates and keep bond yields low by exchanging short-term bonds for longer-term ones. In June of 2012 the Fed decided to extend this program as a way to offset the impending fiscal cliff—when Bush-era tax cuts come to an end and financial woes are expected to worsen—that economists are warning about. "The Federal Reserve will expand its Operation Twist program to extend the maturities of assets on its balance sheet and said it stands ready to take further action to put unemployed Americans back to work," read a June 2012 story in *Bloomberg*. Opponents of Operation Twist claim that it leads to inflation and doesn't produce the intended effects that the Fed claims it has. The media is credited with coining this term to describe the program, which dates back to the 1960s, since the effect it is supposed to have on the yield curve (a chart of interest rates meant to illustrate economic conditions) twists it from arching upward into sloping down. When the Fed previously tried a program like this in the 1960s, Chubby Checker could be heard singing about "The Twist" on the radio and the name stuck.

**RELATED WORDS:** fiscal cliff, taxmageddon

**HOW YOU'LL USE IT:** *"I'm really hoping this OPERATION TWIST program works and we can all relax a little bit about our finances for next year."*

## Orion Capsule *(uh–RAHY–uhn kap–suhl), noun*

Construction for NASA's Orion Capsule by aerospace and technology company Lockheed Martin is expected to be completed this year, with its first test flight expected to happen in 2014. The Orion Capsule is meant to take astronauts into deep space, beyond the orbit of the Earth and further than we've ever been before. Space is the final frontier, after all, and as such we are always pushing the boundaries further. When Orion finally takes its maiden voyage with passengers aboard, a mission that could last six months, it'll be 2021. Rumor has it that the goal of the mission will be to put astronauts on Mars, rather than just the rovers and robots to which we've grown accustomed. The capsule is a multipurpose crew vehicle (MPCV), which looks similar in shape to the Apollo-era vessels. What sets it apart is its technology. Orion will need to have advanced heat shields to guard against radiation and will require better propulsion methods.

**RELATED WORDS:** deep-space exploration, Kennedy Space Center, Lockheed Martin, Mars, NASA

**HOW YOU'LL USE IT:** *"If they use the ORION CAPSULE to go to Mars, that will make the concept of starting a colony there someday seem more real."*

**P**

## PANSTARRS comet *(PAN-stahrz), noun*

The PANSTARRS comet, real name c/2011 L4 (PANSTARRS), was discovered in 2011 by a team of astronomers at the University of Hawaii and Manoa using the Pan-STARRS telescope. The name of the telescope stands for Panoramic Survey Telescope and Rabid Response System. It is meant to discover asteroids due to reach Earth (or at least come close). Of course, a side effect to this kind of work means the discovery of comets, stars, and more. Discussion about the PANSTARRS comet are particularly timely this year since it's expected to be visible to the naked eye when watching from the Northern Hemisphere in February or March. Due to its almost parabolic orbit, it is unknown when, or if, the comet could return again. According to a story on Space .com, "the comet's clunky moniker is slightly unusual. Comets are usually named after their discoverers, but in this case such a large team of researchers helped spot the icy wanderer that it took the name of the telescope instead."

**RELATED WORDS:** 2012 DA14 Asteroid, B612 Sentinel, gravity tractor, laser ablation

**HOW YOU'LL USE IT:** *"A bunch of us are going to get all bundled up and go outside to look for the PANSTARRS comet."*

## parklet *(PAHRK-let), noun*

A parklet is a curbside parking space in an urban environment that has been transformed into a micropark, complete with benches, bike racks, public art, potted plants, and even tables and chairs. The concept originated in San Francisco as early as 2005, with an official parklet program in the city launching in 2010. San Francisco has more than two dozen parklets to speak of, with more on the way. The trend has since spread to cities like New York; Asheville, North Carolina; and Boston, which has plans to install parklets this spring. A story by Eric Moskowitz in the *Boston Globe* about the parklets called them "petite, three-season patios" and surmised that "it remains to be seen how willingly Bostonians, known for fiercely coveting and protecting their parking spots, receive the parklets." The parklet trend is associated with a growing movement away from cars and toward the use of bicycles and public transportation instead. Parklets are also a kind of urban beautification, and are a place where people can meet and socialize. It has been said that neighborhoods with parklets see an increase in visitors and they can even stimulate commerce among local businesses. The word parklet connotes a diminutive quality about the space by placing the suffix "let" at the end of the word park. A similar practice can be seen with words like booklet, wristlet, starlet, and tartlet.

**RELATED WORDS:** al fresco, micropark, patio, urban planning

**HOW YOU'LL USE IT:** *"I met the coolest girl while I was hanging out at the South End PARKLET over the weekend."*

**Perla gas field** *(purl-ah GAS feeld)*, **noun**

The Perla gas field, which was first discovered in 1976, is located about 50 miles offshore in the Gulf of Venezuela. You can expect to hear about the Perla gas field this year, after construction on the drilling project is finished and the site starts producing gas. Once it goes into production, it is expected to produce natural gas at a rate of 300 million cubic feet per day from a set of wells that connect to four offshore platforms. These will be linked to the onshore processing facility via a gas pipeline. The Perla gas field is owned jointly by the Spanish company Respol and the Italian company Eni. In a story for Dow Jones Newswires that appeared on the website for Fox Business, Kejal Vyas writes, "Respol and Italy's Eni SpA (E) signed a $4.5 billion deal with state oil monopoly Petroleos de Venezuela, or PdVSA, to develop the Perla gas field, which is slated to help Venezuela meet domestic demand and overcome power shortages." Estimates put the total amount of natural gas in the Perla gas field at more than 16.3 trillion cubic feet.

**RELATED WORDS:** fracking, offshore drilling, Orinoco belt, refinery

**HOW YOU'LL USE IT:** *"I heard on the news that they're going to start drilling at the PERLA GAS FIELD this week off the coast of Venezuela."*

**personhood** *(PUR-suhn-hood)*, *noun*

The personhood movement is an anti-abortion effort that seeks to legally define human life as starting with a fertilized egg. The movement is headed by Keith Mason and his wife Jennifer through the nonprofit group Personhood USA. Amanda Marcotte writes in a post on Slate that "the primary focus of the personhood movement is to ban abortion, but it's also expected to have the effect of banning women from doing anything that could, in theory, damage a fertilized egg." This means no forms of birth control that would stop a fertilized egg from implanting—like the copper IUD and the morning-after pill. The personhood movement has been around since the 1970s but has been gaining traction as of late under the direction of the Masons, who have been working hard to put personhood on ballots in states across the country. Keith Mason and the personhood movement were profiled in *Newsweek* in the summer of 2012 as issues of women's reproductive rights became a hot-button issue for the impending election. Opponents of the personhood movement assert that defining life in such a way could have dire consequences for women's rights. Aside from the pro-life argument there is also the concern that defining an embryo as a person could make a woman culpable in the event of miscarriage.

**RELATED WORDS:** abortion, gendercide

**HOW YOU'LL USE IT:** *"You better believe that PERSONHOOD will spur a heated debated in our state if it ever goes on the ballot."*

**pexted** *(PEKS-ted), verb*

Pexted is when a man sends someone a photo of his penis via text message. The use of smartphones that include camera technology has enabled this behavior, which can either be unwanted, as in "Ew gross, that guy I started dating just pexted me," or wanted, as in "My boyfriend has been pexting me while he's away on business as a way to sort of spice things up." From time to time the media will report on an incident involving high-profile pexting, like when former professional football player Brett Favre sent some X-rated photos to a sport reporter named Jenn Sterger in 2008 or when professional basketball player Ron Artest pexted an undergrad he'd befriended via Twitter in 2011. More broadly referred to as sexting, which is the transmission of any racy photos or sexually suggestive messages sent via text, the word pexted can be largely attributed to the ABC sitcom *Happy Endings*, which in 2011 included a brief plot point involving the unlucky-in-love character Penny (played by Casey Wilson) whose date started pexting her while she was hanging out with friends. Expect to see the use of pexted increase as new cases arise and as a way to differentiate from text-based sexting.

**RELATED WORDS:** sext

**HOW YOU'LL USE IT:** *"I had to break up with that guy after he wouldn't stop PEXTING me in the middle of the night."*

**phantom vibrations** *(fan-tuhm vahy-BREY-shuhnz)*, **noun**

If you've ever pulled your phone out of your pocket, walked over to it from across the room, or glanced down at it because you thought it was ringing only to be wrong, you've experienced phantom vibrations. We have become increasingly reliant on our cell phones in the last ten years—some may even use the word "obsessed"—to the point that we sometimes imagine that the phone is ringing when it's not. Phantom phone vibrations have also been referred to as ringxiety, since the imagined vibration (or ring) is often associated with an anxiety about not hearing one's phone and thus missing out on a call, text message, email, or social-media update. In this way, phantom vibrations are closely related to FOMO, the "fear of missing out." The science behind this phenomenon is still slim, as only a handful of studies have been conducted on the matter. Expect to hear more about phantom vibrations in the coming years as we get more scientific data on the concept.

**RELATED WORDS:** FOMO, phantom ringing, phantom text syndrome, ringxiety

**HOW YOU'LL USE IT:** *"I swear I just heard my phone going off but then I checked it and it wasn't—creepy PHANTOM VIBRATIONS again!"*

**P**

## Phoenix *(FEE-niks), noun*

Circling the Earth some 22,000 miles above the ground is a graveyard of nearly 140 satellites that could soon be getting new life. The Phoenix satellite program, which is an initiative of the Department of Defense, is working on ways to repurpose and reboot these retired crafts. DARPA—the Defense Advancement Research Projects Agency—is the Department of Defense agency that is overseeing some twenty various government laboratories and companies working on ways to make Phoenix happen. The name Phoenix is no doubt a reference to the mythical bird that rose from the ashes to get another chance at life. The Phoenix satellite program has $44.5 million in funding for 2013. It is the hope of DARPA project manager Dave Barnhart that they will start to convert the old satellites in 2015 through a process that involves launching a mini-satellite, called a satlet, into space to meet up with a robotic servicing satellite, called a tender, which will connect it to a retired satellite. Reusing old satellites will help cut down on costs, as well as preserve some of the pieces of machinery that are already floating in space.

**RELATED WORDS:** DARPA, orbit, satlet, solar storm, tender

**HOW YOU'LL USE IT:** *"I'm glad that the PHOENIX satellite program is figuring out new ways to use old satellites that are orbiting the Earth rather than just hurtling new ones up there all the time and adding to all the stuff that's floating around."*

## pink slime *(pingk SLAHYM), noun*

Pink slime is ground-up beef parts treated with ammonia. Beef manufacturers and consumer-safety officials maintain that it is safe and edible and that pink slime is a derogatory term putting a negative spin on something that has been used in ground beef products since the 1990s. On the other hand, opponents of pink slime are grossed out and concerted about their food safety. The phrase pink slime was first used in 2002 when a scientist at the Agriculture Department named Gerald Zernstein used it in an internal email to refer to what beef companies calls "lean finely textured beef," or LFTB. The point of Zernstein's email was to convey that this ingredient should not actually be considered ground beef. The *New York Times* then published the term in 2009, in Michael Moss's Pulitzer Prize–winning series about food safety in the beef industry. The most notable use though is by celebrity chef Jamie Oliver, who graphically illustrated the components of pink slime on his show *Jamie Oliver's Food Revolution*. This year will bring further discussions about pink slime as food manufacturers such as Beef Products Inc. have been forced to scale back on production and cut workers in the wake of the bad PR. You'll also hear talk of pink slime as school districts grapple with whether or not to keep products containing pink slime on the menu.

**RELATED WORDS:** farm to fork, Jamie Oliver, *The Jungle*, lean finely textured beef, pink menace

**HOW YOU'LL USE IT:** *"I've decided to start grinding my own meat so that I can be sure I'm not eating anything with PINK SLIME in it."*

**P**

**plastic bag ban** *(plas-tik BAG ban)*, **noun**

Expect to hear a lot of coverage related to plastic bag bans this year as multiple cities will institute rules against the common grocery bag. Leading the trend against plastic bags is the environmentally conscious state of California, where cities like Santa Cruz will be outlawing the use of plastic bags in grocery stores and other retailers. (Places like Portland, Oregon, and parts of North Carolina already forbid the bags.) Customers will also be charged 10 cents per paper bag needed, in an effort to strongly encourage the use of reusable bags. Some have argued that the 10-cent fee is in fact too low, with the San Francisco–based Save the Bag Coalition arguing that the production of paper bags is worse than plastic bags. They have suggested that the fee should be more like 25 cents. Toronto has also announced that it will ban the use of plastic bags outright in 2013 and there have been talks of a similar move in Quebec, though initiatives to curtail plastic bag use there may make a ban unnecessary. Single-use plastic bags continue to be a major source of pollution from industrial nations—it is estimated that Americans go through nearly 100 billion plastic bags every year.

**RELATED WORDS:** e-waste, Great Pacific Garbage Patch, microplastic

**HOW YOU'LL USE IT:** *"I've been trying to remember my reusable bags more when I go to the store so I can get in the habit of it before the PLASTIC BAG BAN goes into effect."*

**Play-A-Grill** *(PLEY-uh-gril)*, *noun*

P

Blinged-out mouthpieces and bejeweled orthodontics became a cultural phenomenon and a display of wealth in the past decade thanks to rappers and hip-hop icons like Paul Wall. Even as recently as 2012, Olympic gold-medal swimmer Ryan Lochte could be seen sporting a sparkling grill on his teeth. The latest thing to jazz up your mouth is the Play-A-Grill, a retainer-like device you put in your mouth that plays MP3s and is controlled by using your tongue. According to a story by Sam Laird on Mashable. com, "the Play-A-Grill uses bone conduction to transmit sound, similar to how cochlear implants work for the hearing impaired." Parsons design student Aisen Chacin developed the prototype for the device, which was covered by tech-savvy media outlets like *Time* magazine in 2012 as the next big thing in music technology. In his prototype, the device is connected via a wire, but Chacin is hoping to evolve the product so that is works via Bluetooth.

**RELATED WORDS:** bling, "Grillz," MP3

**HOW YOU'LL USE IT:** *"The invention of the PLAY-A-GRILL must make someone like grill-fan Ryan Lochte happy, especially if it could work while he's swimming in the pool."*

## -pocalypse *(POK-uh-lips), suffix*

Lifting the ending of the word apocalypse and tacking it on to just about anything creates a word of epic proportions. When the nation first went through high rates of unemployment recently, it wasn't just a job crisis; it was a jobpocalypse or a hiring-pocalypse. In a blog post titled "Twipocalypse Now: Warnings of a Twitter Bubble," Neal Wiser writes, "Does Twitter's growing popularity and the evolution of the Twitterverse, the combined ecosystem of users, third party services and all things Twitter, foreshadow continued success or impending doom?" In other words, whenever something gets too big, too extreme, too close to a massive failure or upheaval, send in the knights of the -pocalypse. When maple-syrup production hit a major setback in the spring of 2012, it was a Maple syrup–pocalypse. Daniel H. Wilson's 2011 *New York Times* bestseller titled *Robopocalypse* capitalized on the fear that one day the very technology we have created is going to turn on us. The movie version of *Robopocalypse* will hit theaters in 2013. The rules for creating words like this are simple: take a short root word or prefix, specifically something that people are fearful about or that is undergoing some strain, and then add "apocalypse" onto the end but leave off the "a." What kind of new words might 2013 bring? If a series of zoos are closed, that would be a zoopocalypse. Prohibition suddenly back? That's a beerpocalypse. You get the idea.

**RELATED WORDS:** -geddon, robopocalypse, Twi-pocalypse

**HOW YOU'LL USE IT:** *"I'm so ready for this jobPOCALYPSE to be over so I can find someplace new to work."*

**Project Glass** *(proj-ekt GLAS), noun*

Google's ambitious, high-tech Project Glass centers on a pair of augmented-reality eye glasses, called Google Glass, which are expected to go on the market sometime in 2013. The futuristic-looking glasses include a mini-screen set just in front of the eye and just above a user's regular field of vision. The Android-powered cloud system provides access to maps, photos, video chats, notifications, and potentially even text messages and more. While wearing the glasses, you can access the content of the device with buttons along the side of the frame. The glasses are meant to free its users from that all-too-common affliction of getting lost in the screen of one's mobile phone, head down and with eyes away from the real world. *Fast Company* writer Kevin Purdy test drove a pair, writing that "the screen where your images, videos, and notifications would go (these pair only showed a pretend friend's live video of fireworks) is quite outside your normal field of vision. It's there, and it's slightly translucent, but unless you very deliberately raise your eyes, it's just a notification, much like those that pile up in the corners of your computer desktop." Developers at Google's annual conference who were willing to shell out $1,500 for a prototype of the glasses will receive them in 2013, with the version for regular consumers expected to come not too long after.

**RELATED WORDS:** cloud computing, Google Glass, Google X Lab

**HOW YOU'LL USE IT:** *"PROJECT GLASS will probably mean fewer people walking into things while looking down at their phones."*

**Qmilch** *(KYOO-milk), noun*

The concept for clothing made out of milk has been around for a while and with varying degrees of success. But a German clothing designer has helped to reboot milk-based textiles and could very well be the person to thank for the next big trend in eco-friendly clothing. The modern incarnation of this idea is called Qmilch and was developed by the biologist-turned-clothing-designer Anke Domaske, who owns the clothing line Mademoiselle Chi Chi. Milch means milk in German while the Q stands for "quality." Domaske reportedly became inspired to develop Qmilch after watching her ailing grandfather suffer from skin ailments, which caused him to be uncomfortable in clothes made from traditional fabrics. Qmilch is made by spinning the protein casein, which is found in milk, into a fiber that feels like silk and can be washed normally. Domaske claims that Qmilch has health benefits for its wearer too, saying that it can help you to regulate your body temperature and blood circulation. Qmilch is environmentally friendly since it's made from a renewable resource, and is relatively fast and cheap to make.

**RELATED WORDS:** green fashion, microbiology, organic

**HOW YOU'LL USE IT:** *"What I love most about QMILCH clothing is that it feels just like silk but can go through the washer and dryer without being ruined."*

**Remote Rover Experiment** *(ri-MOHT ROH-ver ik-SPER-uh-muhnt), noun*

For those people who always dreamed of being astronauts and exploring outer space—but who couldn't quite cut it—there's the Remote Rover Experiment project, which continues the 2013 trend of bringing space exploration to the layperson. With funding from the crowdfunding platform Kickstarter, the project is the brainchild of a group that has dubbed themselves the Part-Time Scientists. Their mission is to put a robotic rover on the moon and compete for the Google Lunar X-Prize. But in order to test their prototype in an environment that simulates lunar conditions, they took to the Internet to seek funding and get people involved. Those who pledged money to the project will be able to test drive the prototype and experience a "moon landing" for themselves. Others donating can also receive their own Remote Rover prototype, which they can assemble and test at home. What they get depends on how much they donated. According to the website for the Part-Time Scientists, their goal is "to return space exploration back into the public's attention, like it was 40 years ago, and demonstrate that nothing is impossible, when one puts one's mind to it."

**RELATED WORDS:** Chang'e-3 mission, Excalibur Almaz, Google Lunar X-Prize, Part-Time Scientists, Space Adventures, telerobotics

**HOW YOU'LL USE IT:** *"It was always my dream as a kid to be an astronaut so I pledged a ton of money to the REMOTE ROVER EXPERIMENT last year on Kickstarter in order to get a chance to see what it would be like to control a lunar robot."*

**R**

### Revolt *(ri-VOHLT), noun*

Rapper, singer, actor, and producer Sean "Diddy" Combs is adding television executive to his long list of titles with the creation of the cable network Revolt. Partnered with Comcast, the content of the network is entertainment driven, with a focus on music. According to the *New York Times*, "Combs will bill Revolt as a music and news television channel influenced by the nonstop chatter of social networking Web sites." Among its competitors are BET and MTV. After much speculation about the network, Combs made the announcement official in 2012 when he posted a 3:25 minute video of himself on YouTube that confirmed the news. Featuring a close-up shot of Combs speaking directly into the camera, the video started with Combs declaring "Special announcement: the revolution will be televised." He went on to say that the Revolt network will be a place for artists "to show your art the way you want to show your art—raw, uncut, uncensored." Combs is already an entrepreneur in his own right, with things like the Sean John clothing line, Ciroc Vodka, and a chain of restaurants called Justin's under his direction. The creation of Revolt will put Combs in the company of other high-profile names who have recently joined the television network business, such as Oprah Winfrey and Ryan Seacrest. Comcast accepted the proposal from Combs as part of its pledge to include more diversity in its network owners.

**RELATED WORDS:** BET, MTV

**HOW YOU'LL USE IT:** *"I'm planning to switch to cable so I can get P-Diddy's REVOLT network in my apartment."*

**root to leaf** *(root too leef)*, *noun*

R

Root-to-leaf eating has its roots in the nose-to-tail movement. Nose-to-tail eating is the practice of preparing and consuming as many parts of an animal as possible. Beyond the traditional meat cuts we're all familiar with, nose-to-tail cooking includes the organs—referred to by foodies as offal—as well as other things that aren't traditionally consumed in our culture like hooves, ears, and the head. The movement, which is most frequently concerned with pigs, is meant to support a more sustainable model of eating, since consuming only a small fraction of the animal leads to more waste. Countless books, restaurants, and advocates for nose-to-tail eating have sprung up in recent years as the concept continues to become more mainstream. Of course, the trend is mostly isolated to restaurants since it takes a degree of difficulty that home cooks just aren't used to—yet. The latest interpretation of this movement, which we'll most likely hear about this year as it picks up steam, is a vegetable interpretation of the nose-to-tail movement called root to leaf. Like nose-to-tail eating, root-to-leaf cooking focuses on using all of the parts of a plant to make a meal. Broccoli stems, turnip greens, chive flowers, and squash blossoms are all examples of oft-overlooked items that are now getting the root-to-leaf treatment.

**RELATED WORDS:** farm to fork, locavore, offal, slow-food movement

**HOW YOU'LL USE IT:** *"I'm trying to do more ROOT-TO-LEAF eating this summer as a way to challenge my cooking skills and cut down on waste."*

**R**

**rotation curation** *(roh–TEY-shuhn kyoor-rey-shuhn), noun*

Rotation curation refers to the serial contributions of a large, diverse group of people to a single social-media stream like Twitter or Facebook, quite literally curating the feed's content on a rotating basis. The term has been gaining traction since 2011 when Sweden began to allow a different citizen each week to operate the country's twitter account @sweden. Not long after, the Netherlands opened up their handle @Netherlanders to the rotation-curation model as well. The idea behind rotation curation is that it promotes tourism and interest to the country while portraying it as both progressive and tech-savvy. Sweden's rotation curation has since launched a flurry of new participants looking to get in on the game, including @TweetWeekUSA, @WeAreAustralia, and handles for more than thirty other countries, regions, and groups. After publicly campaigning via his own Twitter account and segments on his TV show *The Colbert Report*, Stephen Colbert was denied a curation spot for Sweden's Twitter account in the summer of 2012. The biggest risk involved with rotation curation, of course, is that the curators are unfiltered, an issue that Sweden learned all too well when one of their curators began to send some misguided tweets about Jews in June of 2012. Though the concept of rotation curation certainly didn't originate with the Twitter trend, it is what made it more mainstream and will most likely be the source of continued discussions.

**RELATED WORDS:** Facebook, Instagram, Twitter

**HOW YOU'LL USE IT:** *"My company is thinking of rolling out a ROTATION CURATION of its official Twitter handle."*

# STU

**S**

### Sagittarius A* *(saj-i-TAIR-ee-uhs ey-stahr), noun*

Sagittarius A*—yes, that asterisk is pronounced out loud like "star"—is believed to be a supermassive black hole that scientists predict will consume a nearby gas cloud this year. The mass of Sagittarius A*, which is located at the center of the Milky Way galaxy, is estimated to be the equivalent of 4 million suns. Though many believe this dark, massive area at the center of our galaxy is indeed a black hole, astronomers have been hesitant to say for sure since it's so difficult to see. As a way to better observe Sagittarius A*, some astronomers have developed a project called the Event Horizon Telescope. The idea is to combine the strength of many radio observatories worldwide so as to get better measurements on Sagittarius A*. Though it was first discovered by astronomers Bruce Balick and Robert Brown in 1974, the term will no doubt be talked about in 2013 as the time nears for the gas cloud to enter. According to a Space.com story by Clara Moskowitz, "as [Sagittarius A*] falls nearer and nearer, the cloud is expected to heat up and release bright X-ray radiation that should be visible from Earth."

**RELATED WORDS:** black hole, Event Horizon Telescope

**HOW YOU'LL USE IT:** *"When the SAGITTARIUS A* swallows that gas cloud in 2013, I plan to stay up all night with a lawn chair in my backyard so I can see if its visible."*

## Scannebago *(SKAN-uh-bey-goh)*, *noun*

One way to create and contribute to a complex digital-information platform involves an idea known as the Scannebago, a combination of the words scan and Winnebago. The concept is rather elegant in its simplicity: hop in a recreational vehicle—particularly of the Winnebago brand—on a sort of informational road trip that would involve stopping in certain areas to scan documents right on board the vehicle. Stops could include local libraries, historical societies, schools, and more—the idea being that most would rather observe their fragile documents being scanned than send them away or do it themselves. This concept would help information and documents get into the Digital Public Library of America while also helping to facilitate a local digital database of these works and documents. Both the concept and the coining of the term can be attributed to Emily Gore, Florida State University's Associate Dean of Libraries for Digital Scholarship & Technology Services. Expect to hear mention of the Scannebago concept as efforts to digitize public documents are increased with the expected launch of the Digital Public Library of America in 2013.

**RELATED WORDS:** Digital Public Library of America, Project Gutenberg, Winnebago

**HOW YOU'LL USE IT:** *"If the SCANNEBAGO becomes real, I would love for it to visit the town I grew up in and digitize all of the historical documents from when the town was settled in the 1700s."*

**S**

**Sensastep** *(SENS-uh-step)*, **noun**

An accident that damaged most of the nerves below his left knee led former sea captain Jon Christiansen to call on a couple of friends to help him walk. Though he could walk with a cane, Christiansen's injury made it hard for him to know when his foot was hitting the ground. With the help of his engineer pals Richard Haselhurst and Steve Willens, he developed Sensastep, aptly named since it senses when you step. The device has three distinct parts: an insole laden with pressure-sensitive sensors that goes under the foot, a transmitter that is strapped around the ankle, and a receiver connected to the ear. Each time a person puts pressure on their foot, the ankle transmitter sends a signal, which vibrates on the bone behind the ear and stimulates the user's cochlear nerve. Expect to hear more about Sensastep in 2013, as the product will likely go into production and be available for wide use once the inventors work out a licensing partnership with a medical-device company.

**RELATED WORDS:** Ekso, human exoskeleton, physical therapy

**HOW YOU'LL USE IT:** *"After the stroke damaged my aunt's foot, something like the SENSASTEP could really help her to walk and be more mobile again."*

**Shackleton Epic** *(SHAK-uhl-tuhn ep-ik),* **noun**

S

The Shackleton Epic is the 2013 recreation of a legendary trek by a polar explorer in 1915. Nearly 100 years ago, an explorer named Ernest Shackleton and his crew became stranded aboard their ship, the *Endurance*, during a trip through Antarctica. The ship became trapped in the ice, forcing the crew to stay there for months trying to wait it out. When the ship began to take on water, they set up camp on the ice and took to lifeboats before landing on Elephant Island. From there, Shackleton and some of his crew took a small boat 800 miles to South Georgia Island, which they crossed on foot—32 miles in 36 hours—before finally retrieving help. Everyone was rescued. This epic trek will be recreated for the first time in 2013 by Australian adventurer Tim Jarvis, who will be sailing aboard a replica of the *Endurance* called the Alexandra Shackleton, after the explorer's granddaughter. Jarvis told *Outside* magazine in an article about the Shackleton Epic that he wants others to be aware of the climate change and environmental challenges in Antarctica, as well as inspire a new generation of people through the story of Shackleton's courage and leadership. Jarvis plans to use period gear and will film the voyage for a documentary.

**RELATED WORDS:** Alexandra Shackleton, Ernest Shackleton, Tim Jarvis

**HOW YOU'LL USE IT:** *"The explorer Tim Jarvis and his team are apparently wearing period clothing during the SHACKLETON EPIC—aren't they going to be freezing?"*

**S**

### showrooming *(SHOH-room-ing), noun*

Showrooming is the act of window shopping at a regular brick-and-mortar store but with the intention of buying online for a better price—either once home or while still in the store from a mobile device. Showrooming is a legitimate concern to both small shops like bookstores and big ones like Best Buy and Target, who all offer products that can be easily found online for much cheaper prices. According to the research firm Gartner, mobile payments for 2012 were on target to surpass 2011 by nearly $65 billion. A story by Bob Greene on CNN.com explains showrooming like this: "People will come into stores, look around, stop at items they particularly like—and instead of carrying them to the cash register, will take photos of them, or type a description into their smartphones." As the showrooming problem persists, stores like Best Buy are already investigating ways to thwart the practice, like including extra training with their employees as a way to maintain the customer-service experience or using lasers to interrupt in-store showrooming from mobile phones. The word showrooming means, quite literally, to treat a store like a showroom.

**RELATED WORDS:** cash mob, mobile payments, window shopping

**HOW YOU'LL USE IT:** *"I am actively trying to curb my SHOWROOMING habits by just buying something in the store when I see it."*

**Sky City** *(skahy SIT-ee)*, *noun*

S

It usually takes years of construction to build something like an enormous skyscraper, but the China-based Broad Sustainable Building has plans to build the world's tallest building, called Sky City, in just three months' time. The ambitious construction project comes shortly after the company erected a thirty-story hotel in just fifteen days. Sky City, which will be built in Changsa, China, and designed by a Dubai-based architect, will reportedly top out at 220 stories, or 838 meters. The construction of Sky City, or Sky City One, which it has also been called, would mean that Dubai's Burj Khalifa was no longer the world's tallest building. Sky City is meant to support a new era of urban living and would include 104 elevators and space for more than 100,000 residents. There will also be restaurants, offices, a hotel, hospital, school, and retail shops all housed within the building, making it quite literally a city in the sky. The cost of this multiuse space is expected to be about $628 million. Erecting a structure as quickly as the planners have proposed is accomplished by using factory-made, prefabricated construction materials that allow builders to drastically speed up the construction project. If all goes as planned, Sky City will be completed in January of 2013.

**RELATED WORDS:** Burj Khalifa, Sky City One

**HOW YOU'LL USE IT:** *"If I lived in SKY CITY, I don't know if I'd ever leave—it's like being on a cruise ship with everything you need right there."*

## SkyVue *(SKAHY-vyoo), noun*

S

By 2013, there will be a new addition to the Las Vegas skyline. The focal point of developer Howard Bulloch's SkyVue will be the 500-foot Ferris wheel, which aims to be the world's third largest behind the Singapore Flyer and the Star of Nanchang. The SkyVue wheel will include thirty-two glass-walled gondolas—each can fit up to twenty-four people inside—which will afford visitors panoramic views of Sin City. It will also have two 50,000-square-foot LED screens, which can be used as advertising space. Even the name of the wheel is for sale, according to a story by Robert Klara in *Adweek*. In addition to the wheel, there will also be a two-story complex below which will become a shopping and dining area. It will also include a convention facility and spaces for concerts and sporting events. Many expect that SkyVue, which is located right across from Mandalay Bay, will have a revitalizing effect on Las Vegas and will become a major tourist attraction for the city. The construction of SkyVue is expected to create nearly 700 jobs. Once it's finished, SkyVue is expected to hire some 500 employees. The official opening date for SkyVue is projected for July 4, 2013.

**RELATED WORDS:** Ferris wheel, Sin City

**HOW YOU'LL USE IT:** *"I really want to take a trip to Las Vegas so I can ride in the SKYVUE observation wheel."*

## slacktivist *(SLAK-tuh-vist), noun*

In a piece for *Forbes* in 2012, writer Erik Kain made a prediction that *Time*'s "Person of the Year" would go to the slacktivist, following past picks like "You" in 2006 and "Protester" in 2011. A slacktivist is something of an armchair activist—a person who takes a stand on social issues by doing something small like retweeting or changing out their profile picture to express solidarity. The term is not something new—it was coined in 1995 by Dwight Ozard and Fred Clark—but it's beginning to have its time in the spotlight due to the rise in social media. The word, a combination of slacker and activism, may come across as derogatory, with the connotation that the slacktivist isn't doing enough by way of furthering the cause. In fact, the idea of the slacktivist has evolved over time to mean someone who is supporting social causes or voicing their opinions from the comfort of their own couch and not out in the streets like a traditional activist. In that same *Forbes* piece, Kain writes, "The information age, after all, is the drift toward information-as-currency. The flattening of institutions and the way that social media and information technology are so rapidly changing the power playing field in both private and public organizations is truly remarkable. At the heart of this change is the slacktivist."

**RELATED WORDS:** Facebook, protest, Twitter

**HOW YOU'LL USE IT:** *"My sister is such a SLACKTIVIST, she just retweets everything that other people say about environmental issues but doesn't ever voice her own opinion outright."*

## smart clothes *(SMAHRT klohz)*, *noun*

**S**

Supercapacitor technology related to energy storage in things like batteries, electric cars, and even flashlights is growing and its applications are leading to the development of things like smart clothes. Smart clothes relate to a new product field wherein fabric can be treated in such a way as to have electronic properties that are either woven into it or are the result of a special treatment. According to a 2010 study by Innovative Research and Products Inc., the market for this technology is expected to grow through 2013 and into 2014. A pair of engineers, named Xiaodong Li and Lihong Bao at the University of South Carolina have figured out a way to transform an ordinary cotton T-shirt into a device that can be used to charge the batteries in smartphones, tablets, and other media devices. Meanwhile, a lab in Canada has figured out a way to transform electric parts into something soft and stretchy that can be woven into fabrics. Experts predict that we will see this smart-clothes technology take over couches and car upholstery, allowing you to adjust the temperature or even the audio of the TV simply by swiping the fabric.

**RELATED WORDS:** graphene, supercapacitor

**HOW YOU'LL USE IT:** *"I can't wait for SMART CLOTHES to go on the market so I can charge my cell phone simply by wearing a T-shirt."*

**smart grid** *(SMAHRT grid)*, *noun*

S

The use of the term smart grid, which has been around for years, is on the rise now that new and alternative approaches to energy use are being explored. A smart grid is a way to conserve resources and save money—it often allows users to control their energy use remotely via a cell phone or computer. An explanation of smart-grid technology by David Biello in *Scientific American* says that "it's the telecommunications and information technology industries applying their innovations to the infrastructure that made computers possible, in large part, or overlaying the utility infrastructure with communications and control systems that will allow energy technology to be more productive." In short: it's a way to control our energy systems in a smarter, more effective way. This technology will be made available to people living in the Phillippines in 2013. Introduced by the Manila Electric Co.—also known as Meralco—participation will require a subscription to the service, plus the installation of a smart meter and specialty plugs and controls. Skeptics of the smart-grid system often cite high startup costs as one of its drawbacks. Supporters, however, maintain that we are long overdue for a technological overhaul of our energy delivery methods and that it would be an investment well worth making.

**RELATED WORDS:** Meralco, smart meters

**HOW YOU'LL USE IT:** *"When I buy my own house I want it to be totally decked out with SMART GRID technology so I can control my energy use from wherever I am."*

**S**

**smishing** *(SMISH-ing), noun*

The concept of phishing, which has been in practice for many years, refers to a kind of tactical fraud where one party tries to elicit information, and then ultimately money, from another party by pretending to be a trusted source. Phishing can also refer to an attempt to access things like passwords, usernames, and personal identification numbers. The main tactic used is often the mimicry of an institution like a bank or credit card company. Smishing is the same thing, except it occurs via text message or SMS. SMS stands for "short message service" and the word smishing comes from joining the acronym SMS with phishing. This form of identity theft often involves the promise of a fake prize, with a link to more information. Instead of getting a prize though, clicking on that link provides a scam artist with access to your phone and all of the information that's associated with it. Other times, smishing will involve a serious-sounding claim from the bank that there's an issue with your card. In order to "activate" your card, they'll give you a number to call. The number will send you to a recording where you'll be asked personal information. One thing that helps to detect smishing is that the source number that has sent the SMS won't have enough digits to be an actual phone number, since the source of the smishing is usually an email address.

**RELATED WORDS:** phishing, skimming, SMS

**HOW YOU'LL USE IT:** *"I keep getting these annoying SMISHING messages on my phone—little do they know, I'm not going to fall for it."*

## solar storm *(SOH-ler stawrm)*, **noun**

S

During a solar storm, solar flares from the sun release charged particles that interact with the Earth's magnetic field. The sun goes through fairly regular cycles of this behavior, and many scientists are predicting that 2013 will see a surge in its solar flares during Solar Cycle 24. The last time a storm of a similar size occurred was in 1859 and witnesses observed fires in telegraph offices, electrified cables, and extraordinarily bright Northern Lights. Some have estimated that damages incurred by a similarly sized solar storm could be greater than $1 or $2 trillion. A large solar storm could adversely affect things like power grids, transformers, and satellite communications. Even cell phones, GPS units, and air traffic could be affected. What's important to note, though, is that scientists have worked hard to create safeguards and precautions to put in place during a solar storm—so despite all the warnings, if our systems have been successfully protected, it's possible we may not even notice when the solar storm is occurring. The 2013 event is expected to happen sometime in the spring, with some experts getting even more specific by claiming that it will occur in May.

**RELATED WORDS:** Carrington flare, geomagnetic storm, Solar Cycle 24, solar flare

**HOW YOU'LL USE IT:** *"Man, wouldn't it be bad if that big SOLAR STORM happened during our road trip and our GPS and all our cell phones stopped working?"*

## Sorkinism *(SOHR-kin-iz-uhmz), noun*

Screenwriter and producer Aaron Sorkin is widely known for his use of dazzling, rapid-fire dialogue in TV shows like *Sports Night*, *The West Wing*, and *Studio 60 on the Sunset Strip*. Some of his stand-out movies include *A Few Good Men*, *Moneyball*, and *The Social Network*, for which he won an Academy Award for Best Adapted Screenplay. However, the dialogue of biting zingers and banter that Sorkin is known for is in fact often recycled from one Sorkin project to another—and by the man himself. A video created by editor Kevin Porter that circulated around the Internet during the summer of 2012 illustrated this pattern. The video, "Sorkinisms: A Supercut," strings together those precise turns of phrase that the writer often repeats. Among the Sorkinisms on display are various reincarnations of the phrases "I'm really quite something," and "I'm not other people." Sorkin has a variety of upcoming new projects in the works, such as a musical about Houdini due out this year and a biopic about Apple cofounder Steve Jobs. We'll have to see which Sorkinisms pop up in those.

**RELATED WORDS:** self-plagiarism

**HOW YOU'LL USE IT:** *"Aaron Sorkin's show* The Newsroom *is chock full of SORKINISMS."*

**Space Adventures** *(SPEYS ad-VEN-cherz), noun*

The United States–based space-tourism company Space Adventures is ramping up its efforts for a civilian fly-by of the moon as early as 2015. The price of a ticket aboard this week-long flight that will travel a quarter of a million miles is a cool $150 million. The price of the ticket also includes months of training leading up to the lunar mission, which will use a Russian rocket to propel travelers past the moon. According to a *Popular Science* article by Clay Dillow, the trip could go a little something like this: "aboard a three-seat Russian Soyuz spacecraft (the third seat is for a Russian mission commander), the tourists would launch into orbit where they would rendezvous with a separately launched unmanned rocket, which would jet them the rest of the way to the moon." Space Adventures, which was founded by Eric C. Anderson in 1998, has already launched a number of paying customers into orbit. It will be competing with the British-run space company Excalibur Almaz to be the first to send a civilian on a trip to the moon. The coming years will see growing coverage of the space-tourism trend as private companies seek to explore more of this final frontier. Sir Richard Branson's space-tourism company Virgin Galactic has plans for a mission into the Earth's orbit in 2013.

**RELATED WORDS:** Excalibur Almaz, Google Lunar X-Prize, space tourism, Virgin Galactic

**HOW YOU'LL USE IT:** *"I get so nervous flying on a regular airplane—I can't imagine how nerve-wracking it would be to fly to the moon with SPACE ADVENTURES."*

## SpaceX *(speys-EKS), noun*

Privately owned SpaceX, which is short for Space Exploration Technologies Corporation, launched its Falcon 9 rocket into space in the spring of 2012, carrying the unmanned Dragon capsule. Dragon successfully docked with the International Space Station to deliver supplies before making it home again. It was the first time a privately owned, operated, and funded ship ever docked with the space station. Since then, SpaceX has been contracted for a dozen more of these cargo missions. SpaceX also has plans to launch a geostationary satellite into orbit in 2013, put a full crew of people into space in 2015, and embark on a mission to Mars in 2018. The company was founded in 2002 by multibillionaire Elon Musk, cofounder of PayPal and Tesla Motors. In a May 2012 article in the *New York Times*, Kenneth Change writes that Musk "is trying to show the world that a determined entrepreneur can start a rocket company from scratch and, a decade later, end up doing a job that has until now been the exclusive province of federal governments." In 2013, we can expect to see continued expansion into space by privately owned companies such as this. Sir Richard Branson's Virgin Galactic is another example as the company has plans to launch its first batch of space tourists off of Earth in 2013.

**RELATED WORDS:** Dragon, Falcon 9, space tourism, Virgin Galactic

**HOW YOU'LL USE IT:** *"All these private companies like SPACEX and Virgin Galactic are creating the space race of our generation."*

**spoofing** *(SPOOF-ing), verb*

S

Spoofing is a tactic used by hackers where a signal sent to a drone mimics the one that is sent to the vehicle's own GPS, allowing the hacker to take control of the vehicle. This type of technology is a big concern to the Department of Homeland Security and researching and developing ways to inhibit spoofing attacks will remain a top priority in the coming years. On June 19, 2012, a team of researchers out of the University of Texas used spoofing tactics to hijack a drone. According to a story by Lorenzo Franceschi-Bicchierai on Wired.com, "some of the drone manufacturers have their own systems to counter spoofing attacks, but others either think this is not their job, are not worried at all, or were completely taken by surprise." The word spoofing is only recently being applied to the practice of overtaking drones. The original meaning of the word simply referred to a network security breach, though the methods and tactics to accomplish regular spoofing and drone spoofing are very similar.

**RELATED WORDS:** cyberattack, drone, hacker, hijack, jamming, smishing

**HOW YOU'LL USE IT:** *"The fact that a bunch of college kids are SPOOFING a drone and making it fly around makes me really nervous."*

S

## strain paint *(STREYN peynt)*, **noun**

Developed by a team of scientists at Rice University, strain paint is a way to easily identify the strain on things like bridges and airplane wings simply by shining a light on it. The paint is infused with carbon nanotubes—each 50,000 times thinner than a strand of human hair—which can express changes in a structure after being exposed to a laser beam. According to a press release on the Rice University website, "nanotube fluorescence shows large, predictable wavelength shifts when the tubes are deformed by tension of compression." This means that anything that's happening to the structure would also show up in the paint. What makes strain paint special is that it provides an early detection for duress on a structure from afar—other methods of achieving the same results require physical contact to measure such a thing. Putting something like strain paint on the market would require more testing first, but with all of the upcoming developments in space exploration and innovative new building structures, its use can't be too far away.

**RELATED WORDS:** Coral World Park, Sky City, SkyVue

**HOW YOU'LL USE IT:** *"I think I would feel a lot better about flying if I knew that airplane technicians were using STRAIN PAINT as a way to make sure everything was working correctly."*

**supersonic inflatable decelerators** *(soo-per-SON-ik in-fley-tuh-buhl dee-SEL-uh-reht-toorz)*, **noun**

S

Starting in the summer of 2013 at NASA's Jet Propulsion Laboratory in Pasadena, California, the space agency will begin testing something called supersonic inflatable decelerators for the first time since 1972. The purpose of these devices is to help in the landing of large spacecraft. These devices are essentially large, balloon-like vessels that inflate around a vehicle to slow it down during its reentry to Earth before the deployment of a parachute. The name of the project by which the testing will be conducted is the Low-Density Supersonic Decelerator Project—LDSD for short—and will be led by Mark Adler. According to a story on NASA's website announcing the testing, "planetary landers of tomorrow will require much larger drag devices than any now in use." In turn, the development of these devices could lead to real-world use on flights as early as 2018. The coming years—this year in particular—will see a spike in talk about space exploration as space tourism gains traction with programs like Virgin Galactic and Space Adventures, and as new technologies related to space programs are covered in the news.

**RELATED WORDS:** GAIA Mission, space tourism, SpaceX, Virgin Galactic

**HOW YOU'LL USE IT:** *"I'd love to take a trip out to Pasadena and watch from afar while NASA tests those SUPERSONIC INFLATABLE DECELERATORS, but I'm not sure that would ever be allowed."*

**T**

**Tactus Technology** *(TAK-tuhs tek-NOL-uh-jee)*, *noun*

Tactus Technology is a California-based company that has developed a way to create tactile buttons on touchscreen devices. The name of the technology is also being referred to as Tactus and involves setting a thin layer of contained microfluids on top of a screen, which can pool to create raised buttons for typing when needed. According to a story by Lakshmi Sandhana for *Fast Company*, "once the layer morphs to create the buttons they remain stable and you can type on them for seconds or hours. Triggered to rise or fall by a proximity sensor or software, button shape, size, height, and firmness can be very minutely controlled." Tactus creates these dynamic touchscreens with the feel of a physical touchscreen in an otherwise flat environment. This means that a person can use their touchscreen phone as normal, but when it comes time to dial the phone, use a calculator, or type a text message the buttons appear. The people at Tactus hope to extend this technology beyond smartphones and adapt it to navigation devices, remotes, dashboard controls, and more. Another benefit to this technology is that it allows for the inclusion of the visually impaired, since the increase in flat touchscreen technology has posed a unique set of challenges within that community. Tactus is expected to become available to manufacturers sometime in 2013.

**RELATED WORDS:** microfluids, touchscreen

**HOW YOU'LL USE IT:** *"I can't wait for TACTUS TECHNOLOGY to be available for the iPhone—I hate that I can't feel where the buttons are when I'm typing something."*

**tanorexic** *(tan-uh-REK-sik), adjective*

T

A slang term describing a condition wherein a person is addicted to becoming tan. This can manifest itself through frequent tanning bed use as well as excessive time spent laying outside in the sun. The derivation of the word tanorexic is a play on the word anorexic, an eating disorder and form of body dysmorphic disorder, wherein a person sees themselves as weighing more than they actually do. Likewise, someone who is tanorexic may not be able to see how tan they actually are—they see themselves as pale. The term tanorexic is not something that has been accepted by the medical community, but is instead a word that gets used colloquially to refer to someone who is obsessive about being tan. If you run a search in Google using the term, one of the prevailing images you'll see is the deeply tanned face of Patricia Krentcil, a mom who in 2012 was brought up on child endangerment charges for allegedly bringing her young daughter tanning. Though the word tanorexic was certainly in use before this incident, Krentcil helped make it part of the conversation once again. Expect to hear the word continuing into 2013 as a Chicago ban keeping teenagers under the age of eighteen out of tanning salons, and a similar New Jersey ban for anyone under sixteen, goes into effect.

**RELATED WORDS:** tanorexia

**HOW YOU'LL USE IT:** *"I stopped going tanning because I don't want to end up looking like that TANOREXIC mom on the news."*

**T**

**taxmageddon** *(TAKS-muh-ged-n), noun*

Taxmageddon describes the impending economic culture our nation could be facing this year. The word is created by lifting the end of the doomsday word Armageddon, which is then affixed to the word tax. (A similar practice is used for words like foodmageddon and Eurogeddon.) The uncertain economic climate that Americans faced last year is expected to continue well into the new year with the end of Bush-era tax cuts (which will hit top wage earners the hardest) and the payroll tax cut (which will hit the middle class and lower-income workers the most). "The potential shock to the nation's pocketbook is so enormous, congressional aides have dubbed it 'Taxmageddon,'" writes Lori Montgomery in the *Washington Post*. "Some economists say it could push the fragile U.S. economy back into recession, particularly if automatic cuts to federal agencies, also set for January, are permitted to take effect." If all proceeds as planned, this could also be the biggest tax hike that Americans have ever experienced. While this word is being thrown around a lot by the media, we must also remember that what it describes is not a guarantee. Some maintain that the risks and potential issues we face this year can be prevented. The theory is that the mere threat of a taxmageddon will jumpstart better programs and deficit solutions so that a fiscal cliff and a further financial crises can be avoided.

**RELATED WORDS:** -mageddon, -pocalypse, fiscal cliff, Operation Twist

**HOW YOU'LL USE IT:** *"Can someone tell me if TAXMAGEDDON means I should just start stockpiling cash under my mattress?"*

**therapeutic hypothermia** *(ther-uh-PYOO-tik hahy-puh-THUR-mee-uh)*, **noun**

An emergency medical treatment known as therapeutic hypothermia, in which the core temperature of a patient's body is cooled to 90 degrees, is gaining traction in 2013 as more reports surface of its successful use. Cooling the body as a method for medical treatment has been around for thousands of years, perhaps even originating with Hippocrates in ancient Greece. Therapeutic hypothermia is most commonly used with patients suffering from cardiac arrest, and was publicized as a method for treatment in a 2011 article in the *Annals of Neurology*. A study released in 2012 in the journal *Circulation* substantiated claims that the practice helps to preserve brain function. The systematic cooling of the human body makes brain and organ functions slow down and has been found to ward off neurological damage during a traumatic event. A pregnant woman who went into cardiac arrest before her water broke was treated with therapeutic hypothermia in 2010 after her baby was delivered via an emergency cesarean section. After 40 minutes of cardiac arrest, the woman was revived. Years later, news outlets are still checking in on the pair—and they're both doing just fine. Expect to hear more about therapeutic hypothermia as reports like this increase in 2013.

**RELATED WORDS:** cardio cooling, Hippocrates

**HOW YOU'LL USE IT:** *"A guy at my office made a full recovery after he had a heart attack—the doctors treated him with THERAPEUTIC HYPOTHERMIA when his heart stopped."*

**T**

**thinspiration** *(thin-spuh-REY-shuhn)*, *noun*

Thinspiration refers to any photos, particularly found online, that have been shared via social media as an inspiration for weight loss. The term, which is sometimes shortened to thinspo, carries derogatory connotations along with it as thinspirational sites, posts, and boards are often associated with a pro–eating disorder mindset. The word is a portmanteau of thin and inspiration. In 2012, social-networking sites like Tumblr, Instagram, and Pinterest all sought to rid their pages of thinspiration photos by banning the content. When Instagram launched its ban on the photos, it also made hashtags like #thinspiration, #proanorexia, and #probulimia unsearchable. Websites dedicated to this idea have in fact been around for years. In addition to featuring photos of skinny women as inspiration, they also post photos of women they deem to be fat, especially celebrities. Such content is often accompanied by derisive and derogatory remarks. (Supermodel Kate Upton was among those widely criticized by thinspiration websites in 2012.) In 2013, the trend will certainly continue as websites continue to grapple with the content and as new coverage of their efforts persists. Those opposed to the banning of the content say that it will not succeed in stopping thinspiration-themed posts, but will instead simply relocate it to different places.

**RELATED WORDS:** Instagram, Pinterest, pro-ana, pro-anorexia, thinspo, pro-anorexia, Tumblr

**HOW YOU'LL USE IT:** *"Even though Pinterest banned all THINSPIRATION boards, I still see stuff like that pop up on my feed all the time."*

**tidal energy** *(TAHY-dl EN-er-jee), noun*

A likely source of job creation in the foreseeable future, tidal energy harnesses the natural ebb and flow of tidal patterns as a way to generate electricity. The first project in the United States to utilize tidal energy was dedicated in Maine on July 24, 2012. Built by Ocean Renewable Power Co. of Portland, Maine, the project involved submerging twenty underwater turbines to create a tidal generator on the ocean floor. Despite the fact that tidal energy is just starting to come into vogue in the United States, it's not actually all that new of an idea. Dating back to the eighteenth century, Maine has used wheels in the water to power various machines. According to a story by Erin Ailworth in the *Boston Globe*, "the Energy Department estimates wave and tidal currents on the nation's coasts have the potential to generate up to roughly one-third of the nation's total annual electricity consumption." Job creation, especially related to the clean energy field, will be discussed through 2013 and into the coming years as ocean energy options continue to be explored.

**RELATED WORDS:** Cape Wind, clean energy, ocean energy, smart grid, solar power, wind power

**HOW YOU'LL USE IT:** *"My brother decided to move up to Maine to see if he could get some work on that new TIDAL ENERGY project they have."*

**-tourism** *(TOOR-iz-uhm), suffix*

If tourism is the act of traveling simply for pleasure, then the added prefix in front of that word connotes traveling with a very specific purpose. Voluntourism, for one, is a trip that includes a philanthropic side. Sure, it's always fun to go somewhere new, but this kind of travel involves volunteer or charity work as well. From coaching soccer in Argentina to teaching English as a second language in Asia, voluntourism is certainly not for those looking to relax. Expanding on this idea are words like socially responsible ecotourism, which is traveling to a remote area to enjoy, protect, and bring awareness to endangered wildlife. A major component of ecotourism is about having a low impact on the environment—a "leave-no-trace" mindset—while also promoting conservation for the area. This can also be called "jungle tourism," since a lot of ecotourists travel to rain forests and other remote jungles. Agritourism, on the other hand, is somewhere in the middle of the two. This trip combines volunteer work with a mind toward sustainability. It is a farm stay or trip to an agricultural area that includes chores like milking cows and picking fruit. Agritourism is also sometimes referred to as "agritainment."

**RELATED WORDS:** agritourism/agritourist, ecotourism/ecotourist, vacationary, voluntourism/voluntourist

**HOW YOU'LL USE IT:** *"I'm planning to use my vacation time to do some volunTOURISM this summer, teaching English in Vietnam."*

## Tupac Katari *(TOO-pahk kuh-TAH-ree), noun*

Named for a Bolivian figure who helped the indigenous people rebel against the Spanish empire in the late-eighteenth century, the Tupac Katari telecommunications satellite will be Bolivia's first once it's set into orbit. The satellite, which is being jointly designed and built in China via a partnership between the Bolivian Space Agency and the Great Wall Industry Corporation of China, is expected to be completed and sent into space sometime in 2013. Part of the partnership with China will also include training for seventy-four members of the Bolivian Space Agency and other specialists from various scientific fields who have been selected from a pool of applicants. The completion of the $300 million project and the subsequent launch of the satellite are expected to have a positive impact on the medical, education, and communications fields in Bolivia. When the satellite is launched, it will be from China's Xichang Satellite Launch Center, and it will communicate with Bolivia via two stations on the ground.

**RELATED WORDS:** 2012 DA14 asteroid, Bolivian Space Agency, China, satellite

**HOW YOU'LL USE IT:** *"Every time I hear about the TUPAC KATARI in the news I think it's a reference to deceased rapper Tupac Shakur and not a Bolivian satellite."*

**U**

## undergrounding *(uhn-der-GROUN-ding), verb*

Undergrounding refers to the practice of burying power lines underground as a way to thwart power outages. It has the added benefit of giving neighborhoods more curb appeal without all the unsightly wires. The concept isn't all that new—many areas already bury some or all of their electrical, cable, and phone lines underground—but is gaining traction due to recent debates about the practice. A pair of July 2012 editorials in *USA Today* has helped to stir up some debate about the pros and cons of undergrounding. Supporters maintain that downed power lines are a major source of frustration for residents following storms and can be a drain on the city as workers try to fix them. The simple solution then, to those who are in favor of the practice, is just to bury them. Opponents of undergrounding say that burying the lines won't magically fix all these issues and that they'll still be susceptible to damage from things like lightning strikes and floods. Expect to hear more about undergrounding as various neighborhoods, towns, and cities continue this debate into 2013 and elect to move ahead with efforts to bury lines, like the town of Naples, Florida, which plans to have their $6.6 million undergrounding efforts completed by the spring.

**RELATED WORDS:** derecho, geoengineering

**HOW YOU'LL USE IT:** *"I'm starting a petition in my town for the UNDERGROUNDING of all our electrical wires because I'm sick of losing power after big storms."*

**Urbano Progresso** *(uhr-BAHN-oh prah-GRES-oh), noun*

U

This year look out for the U.S. release of Kyocera's new cell phone, the Urbano Progresso, which is already available in Japan. The main selling point of the Urbano Progresso is that it's easier to hear someone talking, even if you are speaking to them from a noisy place, because of an innovative new technology called a smart sonic receiver. The super high-tech phone takes advantage of something called tissue conduction to accomplish this. The phone has a ceramic actuator that sends out the sound vibrations through a person's facial tissues rather than the bone conduction method used by some Bluetooth devices. According to a post by Tim Gideon in *Popular Science*, "vibrations move from the actuator through the phone's screen, into the skin and tissue of the face, and on to the eardrum." The Urbano Progresso is the first phone to utilize this tissue-conduction technology, which eliminates the need for the phone's traditional earpiece. Aside from its sound technology, the touchscreen smartphone has an 8.1-megapixel camera, a 4.1-inch screen, and a waterproof casing.

**RELATED WORDS:** bone phone, Facebook phone, Play-A-Grill, smart sonic receiver, tissue conduction

**HOW YOU'LL USE IT:** *"I could really use a phone like the URBANO PROGRESSO when I'm trying to make calls from the noisy train during my commute."*

**U**

## U.S. Ignite *(yoo-es ig-NAHYT)*, *noun*

U.S. Ignite is a government-backed, public-private initiative with support from the National Science Foundation whose goal is to develop a faster broadband network to be used for the development of more sophisticated Internet applications. The creation of a superfast broadband network, called GENI—which stands for Global Environment for Network Innovations—got the go-ahead from President Obama in 2012. GENI will open the door for web developers, who will be granted access to use it as a sort of "sandbox" for new innovations. According to a story by Alex Fitzpatrick on Mashable.com, "the networks are expected to be rolled out in more than twenty-five cities over the next five to six years. Individual developers, startups and major corporations alike will be welcomed to experiment with the networks." By bringing together these big thinkers, the hope is that the new products developed will contribute to the improvement of education, healthcare, public safety, energy, and manufacturing. It is also the mission of U.S. Ignite to innovate new ways to solve problems and create jobs. The National Science Foundation, which has invested $20 million in the project, is also teaming up with the Mozilla Foundation and the Department of Energy to host a $500,000 prize for the development of high-speed apps.

**RELATED WORDS:** GENI, Google Fiber Project, National Science Foundation

**HOW YOU'LL USE IT:** *"I wonder what sort of great innovations will be developed through U.S. IGNITE in the coming years."*

# V
# W
# X
# Y
# Z

**V**

**vacationary** *(vey-KEY-shuhn-air-ee), noun*

Think of a vacationary as a part-time missionary, someone who travels to a poverty-stricken country for the purpose of doing charity and volunteer work but with the added incentive of spreading Christianity. Often comprising teens or young adults just out of college, vacationary groups will do unskilled work, such as painting or building houses, digging irrigation, and so on. In order to go on these trips, participants will often have to raise money—usually several thousand dollars—to fund travel expenses. It is estimated that the vacationary trend produces upward of $5 billion a year, most of which goes to the travel industry—not the communities they're visiting. Rising criticism of the practice has created an added, sometimes derogatory connotation to the word. Critics of vacationaries assert that the work they're doing is merely missionary light, or more severe yet, not missionary work at all. One of the biggest concerns from critics is that during the few days or weeks that vacationaries spend working in the community, they're usurping precious hours that the locals could spend working and getting paid for. Books like Robert D. Lupton's *Toxic Charity* and *When Helping Hurts* by Steve Corbett and Brian Fikkert have explored this growing trend.

**RELATED WORDS:** -tourism, religious tourism, toxic missionary

**HOW YOU'LL USE IT:** *"Our daughter will be in Rwanda for two weeks this summer doing VACATIONARY work with her church."*

## VASIMR (VAZ-im-ehr), noun

The Variable Specific Impulse Magnetoplasma Rocket—VASIMR for short—will reportedly make a preliminary test launch this year if all goes as planned. Developed by former NASA astronaut Franklin Chang-Diaz, the VASIMR is a plasma rocket designed to ionize hydrogen or helium in order to create thrust. Using hydrogen would allow astronauts to actually refuel while out in space since the element can be found throughout our solar system. Longer flights would likely involve the use of nuclear power. When the VASIMR rocket makes its test flight in 2013, it will likely be to the International Space Station, where it will help recalibrate the station's orbit, which can slip over time. According to NASA's website, the craft wouldn't require any additional fuel since it would draw its power from waste hydrogen provided by the station as well as electricity from its solar panels. The VASIMR would be able to travel further distances over a shorter period of time, which means it could hypothetically make the usual six-month trip to Mars in just thirty-nine days. As such, VASIMR is being looked at as our ticket to the establishment of a colony on Mars.

**RELATED WORDS:** HI-SEAS, ISRO Mars Mission, Mars One, MAVEN, SpaceX

**HOW YOU'LL USE IT:** *"The use of VASIMR could mean more distant—and more dangerous—space exploration in the near future."*

**V**

**VB6** *(vee-bee-SIKS), noun*

A 2013 book by the *New York Times* food writer Mark Bittman will have everyone talking about a new weight-loss strategy called VB6, which stands for "vegan before six." The idea is as simple as that: no meat or dairy during the day until after 6 P.M. The concept has been called "selective-veganism" and Bittman claims to have lost upwards of thirty pounds by adhering to this new way of eating. The name of Bittman's forthcoming book, is *VB6: Eat Vegan before 6:00 to Lose Weight and Restore Your Health . . . for Good*. It's not a new thing for Bittman, who spoke about his VB6 eating habits to NPR back in 2009. Bittman says he began following a VB6 diet plan as a way to combat his high cholesterol, high blood sugar, and sleep apnea. Bittman's view was that he could eat all day along-health conscious guidelines, and then indulge a little in the evenings—within reason of course. Though Bittman's been a VB6 believer for years, the concept is still very new to the rest of the general public and his book will no doubt help to make VB6 one of the biggest food trends this year.

**RELATED:** farm to fork, merroir, WikiCells

**HOW YOU'LL USE IT:** *"I'm on a new diet plan called VB6— vegan before 6:00—so no meat or dairy for me at breakfast or lunch."*

**Venus In-Situ Explorer** *(VEE-nuhs in-SEE-tyoo ik-splawr-er),*
*noun*

Advanced space exploration will continue this year, with the planet Venus as the latest to be visited by a craft launched from Earth. Though the idea originated nearly a decade ago, it has been rumored that the Venus In-Situ Explorer—VISE for short—will finally launch its mission sometime in 2013. The purpose of VISE, which is part of NASA's New Frontiers Program, is to collect information about the rock composition and mineralogy on the surface of the planet Venus. Part of the mission will involve drilling into the planet's crust so the samples that are collected will be untouched by the planet's harsh surface conditions. Another reason for the mission is to figure out how the surface of Venus became so volatile in the first place. It will be NASA's first mission to Venus. Past exploratory missions to Venus by the USSR in 1970 and 1983 proved difficult due to the extreme conditions of the planet, with crafts lasting anywhere from two hours to a mere twenty minutes before the planet's atmosphere crushed them. NASA has said that VISE is designed to last four hours before it is destroyed.

**RELATED WORDS:** NASA, New Frontiers Program, VISE

**HOW YOU'LL USE IT:** *"I wonder what sorts of things the VENUS IN-SITU EXPLORER will discover during its mission?"*

**V**

**vertical farming** *(vur-ti-kuhl FAHR-ming), noun*

Vertical farming is an urban agriculture practice that involves growing crops in skyscrapers or other indoor vertical environments, most often by way of artificial light and hydroponics. It's a trend that's been picking up steam for the last few years and is about to hit a tipping point as more projects go from concept to reality and as preexisting vertical farms continue to evolve. According to leading vertical-farming proponent Dickson D. Despommier in a story by Glenn Rifkin in the *New York Times*, a thirty-story vertical farm spanning just one block could produce the same amount of food as 2,400 acres outside. One such vertical farm is The Plant in Chicago, a former pork plant that has been transformed into an indoor farm that is growing greens and mushrooms, baking bread, raising fish, and using the waste of one effort to feed the next. Founded by John Edel in 2010, The Plant expects to add a compost-fueled generator to the operation at some point this year. Expect to hear more about vertical farming in 2013 as more projects continue to develop, such as Vertical Harvest, a three-story hydroponic farm in Jackson Hole, Wyoming, that is expected to begin construction soon.

**RELATED WORDS:** agribusiness, hydroponics, Lufa Farms, urban rooftop farming,

**HOW YOU'LL USE IT:** *"There are so many abandoned buildings in my neighborhood that would be perfect for VERTICAL FARMING."*

**Virgin Galactic** *(VUR-jin guh-LAK-tik), noun*

For every little kid who has ever dreamed of becoming an astronaut and launching into space but became something like an accountant or a tax attorney instead, Virgin Galactic may have the answer. Founded by English billionaire Sir Richard Branson in 2004, Virgin Galactic announced that they may finally be able to take tourists into space this year. Already, some 500 people have purchased tickets for the ride, which checks out at $200,000 a pop—pretty steep for a two-and-a-half-hour suborbital ride, though the trip does include the promise of weightlessness. Some notable names who have signed on to be among the first include Ashton Kutcher and Stephen Hawking, while other celebrities like Brad Pitt, Angelina Jolie, and Tom Hanks are rumored to be on board with the mission as well. When the inaugural flight does launch into space, it will be from the New Mexico desert at the Virgin Spaceport, with Sir Branson himself onboard the SpaceShip Two. The Virgin magnate has also told the press that he intends to have his children on board with him. Former Air Force test pilot Keith Colmer has been hired as the SpaceShip Two's inaugural pilot.

**RELATED WORDS:** -tourism, space tourism, SpaceX

**HOW YOU'LL USE IT:** *"If I ever win the lottery, the first thing I'm going to do is buy a ticket on VIRGIN GALACTIC so I can see what the Earth looks like from space."*

**W**

**weed pass** *(WEED pas)*, **noun**

Amsterdam's famous marijuana-selling coffee shops are facing big changes in 2013 if/when a law banning tourists from purchasing weed goes into effect. Instead of being able to just walk into a shop and buy a pot brownie, only Dutch-born citizens and members of the so-called weed pass system will be allowed to partake. Locals interested in maintaining the Amsterdam lifestyle will be required to become members of a particular coffee shop, which will be limited to 2,000 per shop. The ban comes about at a time when drug tourism is becoming more prevalent in the area. Dealers from nearby countries, like Germany and Belgium, travel to Amsterdam and buy large quantities of marijuana only to resell it back home. Legislators hope that establishing a weed pass will curb this. The fear among those opposed to the weed pass, however, is that tourism could take a financial hit in the coming years, since many travel to Amsterdam to participate recreationally in this European novelty. Locals have also voiced concerns that a ban on tourists will create a rise in petty crime as well as a return to a black-market system with travelers buying from dealers on the street.

**RELATED WORDS:** wiet pass (Dutch spelling)

**HOW YOU'LL USE IT:** *"We decided to cancel our trip to Amsterdam now that they have the WEED PASS—that's really the only reason we were going there."*

## Wheat Belly Diet *(WEET bel-ee dahy-it)*, noun

**W**

Dieting fads come and go fairly regularly, with promises of weight loss simply by following the latest rule of what—and what not—to eat. The latest diet you'll be talking about (or maybe even trying) is the Wheat Belly Diet. Dr. William Davis and his book *Wheat Belly* are to thank for kickstarting this diet, which requires you to forgo all products that contain wheat, as well as prepackaged items that are gluten-free. This means no bread, no pasta, no pastries, no tortillas, or any food with wheat. Even whole-wheat bread and pasta are forbidden. Instead, dieters are to eat meals rich in meat and vegetables as well as eggs, nuts, and fish. Fruit is to be limited since the fructose it contains is a simple carbohydrate. The thought behind Davis's Wheat Belly Diet is that it helps you to lose the "wheat belly," as well as overall weight. His argument in favor of the diet is that we have become addicted to wheat due to a protein called gliaden. Davis maintains that as wheat becomes genetically modified and the amount of gliaden present is increased, so do our cravings. Critics of the Wheat Belly Diet don't dismiss that giving up wheat will lead to weight loss—where there's skepticism is in Davis's logic about genetic modifications causing an addiction to wheat.

**RELATED WORDS:** Atkins Diet, celiac disease, gluten, Paleo Diet

**HOW YOU'LL USE IT:** *"I threw out all the bread, cookies, and other wheat-filled products I had in my house since I plan to start the WHEAT BELLY DIET next week."*

**W**

**white fanging** *(what FANG-ing)*, **verb**

In the finale episode of the first season of the popular Fox comedy *New Girl*, the character Schmidt (played by Max Greenfield) breaks up with his love interest Cece (played by Hannah Simone) because he's afraid she'll never be happy with him in the long term. "Are you White Fanging me?" Cece asks Schmidt, then explains: "*White Fang*? The only book you have on your Kindle. The book you wouldn't stop talking about, and I said, 'Would you please stop talking about *White Fang*' and then you said, 'Someday, I'm gonna do that to somebody." Jack London's famous novel about a wolf-dog hybrid named White Fang actually never includes the exchange *New Girl* is referencing. What the *New Girl* writers were drawing from was in fact the 1991 film adaptation starring Ethan Hawke as Jack Conroy, a gold miner in Canada who befriends the wild dog. In the movie version, Jack sends the dog away, freeing him back into the wilderness. It pains him to do this since the pair have formed quite the bond, but he knows it's the only way the animal can be happy. And thus, the term "white fanging" means just that, to breakup with someone when you feel like you're holding them back.

**RELATED WORDS:** breakups, *New Girl*

**HOW YOU'LL USE IT:** *"He's totally WHITE FANGING me— he said he doesn't think he can give me what I'm looking for."*

**wide-view device** *(WAHYD-vyoo dih-vahys),* **noun**

**W**

Developed by Bionic Vision Australia, a wide-view device is a retinal implant—or a bionic eye—that uses a microchip to transmit messages directly to the brain. Inside the device are ninety-eight electrodes meant to stimulate retinal nerve cells and send messages along the optic nerve so that its user can maneuver within the world and see things like buildings, cars, and park benches. According to Bionic Vision Australia's website, the wide-view device is meant to allow patients "to lead more independent lives." Patients using the wide-view device would wear a pair of glasses that have been outfitted with a tiny camera that captures images and then processes them externally—potentially even via a patient's cell phone. From there, an implanted receiver sends the signals to the retinal implant, and then from the retina and along the optic nerve to the brain's vision-processing centers. Bionic Vision Australia hopes to begin testing this new technology on patients this year. The technology will be taken a step further next year, when scientists plan to debut a second version of the device that would be sophisticated enough to help a person read and recognize faces.

**RELATED WORDS:** bionic

**HOW YOU'LL USE IT:** *"My uncle, who is legally blind, told me he wants to be among the patients who test out that WIDE-VIEW DEVICE when it becomes available."*

**W**

## WikiCells *(WIK-ee selz), noun*

WikiCells refers to both the edible packaging and group of products that use this technology created by a team at Harvard University. The goal of WikiCells, and their sister products like WikiCocktail and Wiki Ice Cream, is to create an alternative to wasteful packaging like paper and plastic. So even if you didn't care to eat the packaging, at the very least you could throw it into the compost bin. At its core, WikiCells are meant as a solution to the packaging problem, but they were also designed to be nutritional, too. The developers behind WikiCells have compared their product to the skin of a grape. "Each WikiCell has a nutritional skin held together by healthy ions like calcium," says the product's website. "Think about the skin of a grape and how it protects the grape itself. This is how a WikiCell works." Inside a WikiCell can be things like ice cream, cheese, yogurt—even beverages. The outside coating is made from organic materials and is meant to be rinsed the same way you'd wash an apple or another piece of fruit before eating it. The brain behind WikiCells belongs to David Edwards, a biomedical engineering professor at Harvard who is also known for creating a caffeine inhaler called Aeroshot and a chocolate inhaler called Le Whif. WikiCells will become available for public consumption in the United States this year after debuting in Paris.

**RELATED WORDS:** AeroShot, Le Whif, WikiCocktail, Wiki Ice Cream

**HOW YOU'LL USE IT:** *"Hey let's try that new restaurant up the street—I hear they have WIKICELLS on the menu."*

**World Baseball Classic** *(world BEYS-bawl KLAS-ik), noun*

W

The international baseball tournament known as the World Baseball Classic (or the WBC, as it's also called) will return this year after a planned four-year hiatus. The first and second World Baseball Classics were held in 2006 and 2009, respectively, with Japan as the winner for each. Following the 2013 series, the event will be held every four years, much like the Olympics. Though Japan is the defending champion, there's been some question about whether or not they'd return to the tournament. Threats from Japan began circulating in 2012 that they'd boycott the tournament unless the players received a bigger cut of the revenue generated from the series. This is actually the second time this happened, as Japan also threatened to sit out the tournament in 2009 for similar reasons. There are sixteen teams expected to compete in total and the series is conducted in a double elimination format. The World Baseball Classic is largely popular in Asia, and hasn't quite received the same following in the United States as it has there. Major League Baseball started the series to create baseball's answer to soccer's World Cup.

**RELATED WORDS:** MLB, WBC, World Cup

**HOW YOU'LL USE IT:** *"I'm going to get some people together this weekend to watch some of the WORLD BASEBALL CLASSIC on my big-screen TV."*

### XCOR Aerospace *(EKS-kor AIR-oh-speys)*, *noun*

Space tourism is becoming real in a very big way in 2013 with things like Space Adventures, Virgin Galactic, and Excalibur Almaz. Add to that list XCOR Aerospace, a California-based space-tourism company with its sights set on taking passengers into space from California this year and from Curaçao in 2014. What sets XCOR apart is that each flight would be able to take off and land on a regular airport runway aboard a suborbital Lynx space plane. There's room for just two on board the Lynx—a pilot and a passenger—and XCOR has said it could run up to four flights a day on the reusable plane. Each trip would carry passengers some sixty-two miles above Earth for $95,000 per ride. Included in the price is also the necessary training that passengers would need to go through before the flight. For comparison, a ticket aboard the Virgin Galactic flight costs approximately $20,000.

**RELATED WORDS:** Excalibur Almaz, Space Adventures, space tourism, SpaceX, Virgin Galactic

**HOW YOU'LL USE IT:** *"I like the idea that XCOR AEROSPACE's flights leave from a conventional runway at the airport—there would be something comforting about that familiar experience before taking off into space."*

### X-points *(EKS-points), noun*

Though the name X-points may conjure up images of wormholes or portals straight out of a science-fiction novel, the actual meaning is much less sci-fi. X-points, which are also referred to as "electron diffusion regions," are in fact portals, but not in the way you'd think. In 2012, NASA announced that they had confirmed the existence of X-points that link the Earth directly to the Sun. NASA-funded researcher Jack Scudder is credited with the discovery. Rather than serving as a mode of hyperfast space travel though, X-points instead help to transfer magnetically charged particles. These particles are the same ones responsible for things like the Northern Lights and geomagnetic storms. Expect to hear more about X-points this year as NASA ramps up efforts to study these phenomena in 2014 during a space mission called Magnetospheric Multiscale Mission (MMS). MMS will launch four crafts in search of the exact location of these X-points, which are difficult to locate because they can open and close, and will attempt to learn more about them.

**RELATED WORDS:** electron diffusion regions, geomagnetic, NASA, wormhole

**HOW YOU'LL USE IT:** *"I know X-POINTS aren't actually wormholes like in the movies, but wouldn't it be crazy if the MMS ships got sucked into them?"*

**Y**

**yogurt wars** *(YOH-gert wawrz), noun*

In the past few years, yogurt has become—dare we say—trendy. Greek yogurt especially, as seen by the soaring popularity of brands like Chobani and Fage. "The Yogurt Wars Will Not End Until Every American Is Eating an All-Yogurt Diet" read a 2012 headline on Gawker.com. Yoplait is the most popular brand of yogurt in the United States, and as such is planning a host of new varieties as a way to compete with the rising Greek-yogurt market this year. Dannon is also getting in on the yogurt wars with its own Greek version, among other things. Food-based competition among brands is nothing new, of course, but it's not all that often you think of yogurt sales as something that is competitive. The concept of a food "war" actually originated in the 1980s with the Cola Wars between the Coca-Cola Company and PepsiCo Incorporated, with the brands going head-to-head by way of pop culture–driven television ads. Expect to hear talk of the yogurt wars as the various brands release new and competitive advertising campaigns. The proliferation of frozen-yogurt chains in major cities across the country—with shops sometimes opening within walking distance of one another—is also often referred to as a yogurt war.

**RELATED WORDS:** Chobani, Cola Wars, Dannon, Fage, Yoplait

**HOW YOU'LL USE IT:** *"YOGURT WARS can happen all they want as long as it means yummy new flavors of Greek yogurt will be on the supermarket shelves."*

## -zilla *(ZIHL-uh), suffix*

**Z**

When tacked on to the end of a noun, the suffix "zilla" transforms its root to mean something overbearing, over the top, and unrelenting. It is someone whose behavior has intensified, whose actions have been taken to extremes. As such, "zilla" can be added to almost any noun to convey this meaning. A bridezilla, for instance, is a woman who has become overly demanding and emotionally charged during the planning of her wedding. (A groomzilla is the male equivalent.) The reality TV show *Bridezillas* capitalizes on both the word and phenomenon by documenting the lives of women in the throes of wedding planning. Seating chart not right? Flowers are late? You better believe these bridezillas will throw a fit about it. Following suit, words like momzilla and promzilla extend beyond the wedding to mean any person who is being difficult or demanding in reference to a particular event or situation. It is someone who uses strong language, bullying, and emotional fits to get what he or she wants. At its heart, these -zilla terms are a reference to Godzilla, the Japanese movie monster known for his mighty roar and a penchant for destroying cities.

**RELATED WORDS:** bridezilla, momzilla, promzilla

**HOW YOU'LL USE IT:** *"I cannot wait for my sister's wedding to be over—she has turned into such a brideZILLA these last couple of weeks."*

## About the Author

Nicole Cammorata is a Boston-based journalist. Her work has appeared in the *Boston Globe*, the *Boston Globe Sunday Magazine*, Boston.com, and *Dazed & Confused* magazine.